Praise for *Field Notes from a Catastrophe*

"Each chapter is part of a larger narrative, a loose travelogue that includes the Alaskan interior, Iceland and the Greenland ice sheet, but, more important, these narrative elements, while drawing us in, always keep a larger purpose in sight—to offer the clearest view yet of the biggest catastrophe we have ever faced."
—**Los Angeles Times**

"Comprehensive and succinct."—**New York Times**

"Elegant."—**Chicago Tribune**

"Kolbert weighs in with a slim but very powerful book."
—**Chicago Sun-Times**

"Essential reading. [Kolbert] is an accomplished writer whose prose is deceptively simple and whose meaning is always clear. Climate change is complex stuff, but she deftly distills the brew to clarity."—**Minneapolis Star Tribune**

"Gripping, well-written."—**Seattle Times**

"Short, readable and scrupulously objective . . . Kolbert's central achievement is to have examined the complex panoply of data surrounding the science of climatology and to have presented it in prose as elegant as the facts themselves are stark."—**Times Literary Supplement**

"Very powerful . . . important."—**Hartford Courant**

"Enlightening."—**Oregonian**

"*Field Notes* is a wonderful read, a superbly crafted, urgently compressed vision of a world spiralling towards destruction. It should be a wake-up call to the world."—**Observer (London)**

"Kolbert is like Matisse, painting an evocative picture with a few deft strokes . . . I recommend [*Field Notes*] to anyone who has a concern for the fate of our planet."—**American Scientist**

"Kolbert, a staff writer for *The New Yorker*, surveyed the world's leading climate scientists, and discovered an alarming unanimity to their message: The world needs to wake up, and fast." —**Wired**

"Let's hope this powerful, clear and important book is not just lightly compared to *Silent Spring*. Let's hope it is this era's galvanizing text."—**Scientific American**

"An elegant ride through the confusing world of climate science. Kolbert takes a John McPhee–style ramble across the world: In Greenland, Iceland, and poor little Shishmaref, she sees the effects of warming firsthand. In Washington, D.C., the former *New York Times* political reporter puts her Beltway savvy to use, revealing that the most climate-change skepticism originates in the deep pockets of oil and coal companies."—**Outside**

"Keenly observed and deeply memorable . . . the picture [Kolbert] draws is compelling—and very scary."—**Seed**

"An extraordinary piece of reporting."—**Grist.org**

"Kolbert forces us to ponder a tragic disconnect: Politicians, the group best positioned to do something about the scientists' warnings, turn out to be the group that's most adamantly ignoring those warnings."—**Austin American-Statesman**

"[Kolbert] traverses the globe to observe first hand the jaw-dropping results of climate changes . . . species going extinct, polar caps melting, water rising in the Netherlands or butterflies in England migrating to higher elevations than ever before." —**Contra Costa Times**

"The hard, cold, sobering facts about global warming and its effects on the environment that sustains us. Kolbert's *Field Notes from a Catastrophe* is nothing less than a *Silent Spring* for our time."—**T.C. Boyle, author of *Talk, Talk***

"Kolbert mesmerizes with her poetic cadence in this riveting view of the apocalypse already upon us."—**Robert F. Kennedy, Jr., author of *Crimes Against Nature***

"Reading *Field Notes from a Catastrophe* during the 2005 hurricane season is what it must have been like to read *Silent Spring* forty years ago. When you put down this book, you'll see the world through different eyes."—**Sylvia Nasar, author of *A Beautiful Mind***

"[I]f you know anyone who still does not understand the reality and the scale of global warming, you will want to give them this book."—**Jonathan Weiner, author of *The Beak of the Finch***

"This country needs more writers like Elizabeth Kolbert." —**Jonathan Franzen, author of *The Corrections***

"Deliver[s] an exceptionally clear picture of how global warming is already on us and what disasters lie in wait should we fail to act."—**Salon***

"The brilliance of *Field Notes from a Catastrophe* flows from Kolbert's gift for making the violence of climate change feel vast yet intimate . . . Her field-journal format adroitly bridges the gulf between professionals and amateurs, giving the writing a conversational tone without compromising the science."—**Slate***

The Prophet of Love

FIELD NOTES

from a

CATASTROPHE

MAN, NATURE,
AND CLIMATE CHANGE

Elizabeth Kolbert

BLOOMSBURY

Published by Bloomsbury USA, New York

All papers used by Bloomsbury USA are natural,
recyclable products made from wood grown in well-managed
forests. The manufacturing processes conform to the
environmental regulations of the country of origin.

The Library of Congress has cataloged the hardcover edition as follows:

Kolbert, Elizabeth.
Field notes from a catastrophe : man, nature, and climate change / Elizabeth
Kolbert.—1st U.S. ed.
p. cm.
Includes bibliographical references and index.
ISBN-13 978-1-59691-125-3
ISBN-10: 1-59691-125-5
1. Global warming. 2. Global temperature changes. 3. Global environmental
change. I. Title.
QC981.8.G56K655 2006
363.738'74—dc22
2005030972

First published in hardcover by Bloomsbury USA in 2006
This paperback edition published in 2007

Paperback ISBN 1-59691-130-1
ISBN-13 978-1-59691-130-7

12 14 16 18 20 19 17 15 13 11

Typeset by Hewer Text UK Ltd, Edinburgh
Printed in the United States of America by Quebecor World Fairfield

To my boys

AUTHOR'S NOTE

The language of science is metric; however, most British and American readers speak—and think—in units like feet, miles, and degrees Fahrenheit. I have used English units where practical and metric units where it seemed clearly more appropriate. For instance, the standard measure of carbon emissions is metric tons. A metric ton weighs 2,205 pounds.

CONTENTS

PREFACE

THERE ISN'T MUCH to do at the Hotel Arctic except watch the icebergs flow by. The hotel is located in the town of Ilulissat, on the west coast of Greenland, four degrees north of the Arctic Circle. The icebergs originate some fifty miles away, at the end of a long and fast-moving ice stream known as the Jakobshavn Isbrae. They drift down a fjord and through a wide-mouthed bay, and, if they last long enough, end up in the North Atlantic. (It is likely that the iceberg encountered by the *Titanic* followed this route.)

To the tourists who visit the Hotel Arctic, the icebergs are a thrilling sight: beautiful and terrible in equal measure. They are a reminder of the immensity of nature and the smallness of man. To the people who spend more time in Ilulissat—native Greenlanders, European tour guides, American scientists—the icebergs have come to acquire a different significance. Since the late 1990s, the Jakobshavn Isbrae has doubled its speed. In the process, the height of the ice stream has been dropping by up to fifty feet a year and the calving front has retreated by several miles. What locals now notice about the icebergs is not their power or immensity—though they are still powerful and immense—but a disquieting diminishment.

"You don't get the big icebergs anymore," an Ilulissat town councilman named Jeremias Jensen told me. We were having coffee on a late-spring afternoon in the Hotel Arctic lobby. Outside, it was foggy and the icebergs seemed to be rising up out of the mist. "It's very strange the last few years; you can see a lot of strange changes."

This is a book about watching the world change. It grew out of three articles that I wrote for the *New Yorker* magazine, which ran in the spring of 2005, and its goal remains much the same as that of the original series: to convey, as vividly as possible, the reality of global warming. The opening chapters are set near or above the Arctic Circle—in Deadhorse, Alaska; in the countryside outside of Reykjavík; at Swiss Camp, a research station on the Greenland ice sheet. I went to these particular places for all the usual journalistic reasons—because someone invited me to tag along on an expedition, because someone let me hitch a ride on a helicopter, because someone sounded interesting over the telephone. The same is true of the choices that were made in subsequent chapters, whether it was a decision to track butterflies in northern England or to visit floating houses in the Netherlands. Such is the impact of global warming that I could have gone to hundreds if not thousands of other places—from Siberia to the Austrian Alps to the Great Barrier Reef to the South African fynbos—to document its effects. These alternate choices would have resulted in an account very different in its details, but not in its conclusions.

Humans aren't the first species to alter the atmosphere; that distinction belongs to early bacteria, which, some two billion years ago, invented photosynthesis. But we are the first species to be in a position to understand what we are doing. Computer models of the earth's climate suggest that a critical threshold is approaching. Crossing over it will be easy, crossing back quite likely impossible. The second part of this book explores the complicated relationship between the science and the politics of global warming, between what we know and what we refuse to know.

My hope is that this book will be read by everyone, by which I mean not only those who follow the latest news about the climate but also those who prefer to skip over it. For better or (mostly) for worse, global warming is all about scale, and the sheer number of figures involved can be daunting. I've tried to offer what is essential without oversimplifying. Similarly, I have tried to keep the discussion of scientific theory to a minimum while offering a full-enough account to convey what is truly at stake.

Part I

NATURE

Chapter 1

SHISHMAREF, ALASKA

THE ALASKAN VILLAGE of Shishmaref sits on an island known as Sarichef, five miles off the coast of the Seward Peninsula. Sarichef is a small island—no more than a quarter of a mile across and two and a half miles long—and Shishmaref is basically the only thing on it. To the north is the Chukchi Sea, and in every other direction lies the Bering Land Bridge National Preserve, which probably ranks as one of the least visited national parks in the country. During the last ice age, the land bridge—exposed by a drop in sea levels of more than three hundred feet—grew to be nearly a thousand miles wide. The preserve occupies that part of it which, after more than ten thousand years of warmth, still remains above water.

Shishmaref (population 591) is an Inupiat village, and it has been inhabited, at least on a seasonal basis, for several centuries. As in many native villages in Alaska, life there combines—often disconcertingly—the very ancient and the totally modern. Almost everyone in Shishmaref still lives off subsistence hunting, primarily for bearded seals but also for walrus, moose, rabbits, and migrating birds. When

I visited the village one day in April, the spring thaw was under way, and the seal-hunting season was about to begin. (Wandering around, I almost tripped over the remnants of the previous year's catch emerging from storage under the snow.) At noon, the village's transportation planner, Tony Weyiouanna, invited me to his house for lunch. In the living room, an enormous television set tuned to the local public-access station was playing a rock soundtrack. Messages like "Happy Birthday to the following elders . . ." kept scrolling across the screen.

Traditionally, the men in Shishmaref hunted for seals by driving out over the sea ice with dogsleds or, more recently, on snowmobiles. After they hauled the seals back to the village, the women would skin and cure them, a process that takes several weeks. In the early 1990s, the hunters began to notice that the sea ice was changing. (Although the claim that the Eskimos have hundreds of words for snow is an exaggeration, the Inupiat make distinctions among many different types of ice, including *sikuliaq*, "young ice," *sarri*, "pack ice," and *tuvaq*, "land-locked ice.") The ice was starting to form later in the fall, and also to break up earlier in the spring. Once, it had been possible to drive out twenty miles; now, by the time the seals arrived, the ice was mushy half that distance from shore. Weyiouanna described it as having the consistency of a "slush puppy." When you encounter it, he said, "your hair starts sticking up. Your eyes are wide open. You can't even blink." It became too dangerous to hunt using snowmobiles, and the men switched to boats.

Soon, the changes in the sea ice brought other problems. At its highest point, Shishmaref is only twenty-two feet above sea level, and the houses, most of which were built by the U.S. government, are small, boxy, and not particularly sturdy-looking. When the Chukchi Sea froze early, the layer of ice protected the village, the way a tarp prevents a swimming pool from getting roiled by the wind. When the sea started to freeze later, Shishmaref became more vulnerable to storm surges. A storm in October 1997 scoured away a hundred-and-twenty-five-foot-wide strip from the town's northern edge; several houses were destroyed, and more than a dozen had to be relocated. During another storm, in October 2001, the village was threatened by twelve-foot waves. In the summer of 2002, residents of Shishmaref voted, a hundred and sixty-one to twenty, to move the entire village to the mainland. In 2004, the U.S. Army Corps of Engineers completed a survey of possible sites. Most of the spots that are being considered for a new village are in areas nearly as remote as Sarichef, with no roads or nearby cities or even settlements. It is estimated that a full relocation would cost the U.S. government $180 million.

People I spoke to in Shishmaref expressed divided emotions about the proposed move. Some worried that, by leaving the tiny island, they would give up their connection to the sea and become lost. "It makes me feel lonely," one woman said. Others seemed excited by the prospect of gaining certain conveniences, like running water, that Shishmaref lacks. Everyone seemed to agree, though, that the village's situation, already dire, was only going to get worse.

Morris Kiyutelluk, who is sixty-five, has lived in Shishmaref almost all his life. (His last name, he told me, means "without a wooden spoon.") I spoke to him while I was hanging around the basement of the village church, which also serves as the unofficial headquarters for a group called the Shishmaref Erosion and Relocation Coalition. "The first time I heard about global warming, I thought, I don't believe those Japanese," Kiyutelluk told me. "Well, they had some good scientists, and it's become true."

The National Academy of Sciences undertook its first major study of global warming in 1979. At that point, climate modeling was still in its infancy, and only a few groups, one led by Syukuro Manabe at the National Oceanic and Atmospheric Administration and another by James Hansen at NASA's Goddard Institute for Space Studies, had considered in any detail the effects of adding carbon dioxide to the atmosphere. Still, the results of their work were alarming enough that President Jimmy Carter called on the academy to investigate. A nine-member panel was appointed. It was led by the distinguished meteorologist Jule Charney, of MIT, who, in the 1940s, had been the first meteorologist to demonstrate that numerical weather forecasting was feasible.

The Ad Hoc Study Group on Carbon Dioxide and Climate, or the Charney panel, as it became known, met for five days at the National Academy of Sciences' summer indeed, entire books have been written just on the history of efforts to draw attention to the problem. (Since the Charney report, the National Academy of Sciences alone has produced nearly two hundred more studies on the subject, including, to name just a few, "Radiative Forcing of Climate Change," "Understanding Climate Change Feedbacks," and "Policy Implications of Greenhouse Warming.") During this same period, worldwide carbon-dioxide emissions have continued to increase, from five billion to seven billion metric tons a year, and the earth's temperature, much as predicted by Manabe's and Hansen's models, has steadily risen. The year 1990 was the warmest year on record until 1991, which was equally hot. Almost every subsequent year has been warmer still. As of this writing, 1998 ranks as the hottest year since the instrumental temperature record began, but it is closely followed by 2002 and 2003, which are tied for second; 2001, which is third; and 2004, which is fourth. Since climate is innately changeable, it's difficult to say when, exactly, in this sequence natural variation could be ruled out as the sole cause. The American Geophysical Union, one of the nation's largest and most respected scientific organizations, decided in 2003 that the matter had been settled. At the group's annual meeting that year, it issued a consensus statement declaring, "Natural influences cannot explain the rapid increase in global near-surface temperatures." As best as can be determined, the world is now warmer than it has been at any point in the last two millennia, and, if current trends continue, by the end of the century it will likely be hotter than at any point in the last two million years.

In the same way that global warming has gradually ceased to be merely a theory, so, too, its impacts are no longer just hypothetical. Nearly every major glacier in the world is shrinking; those in Glacier National Park are retreating so quickly it has been estimated that they will vanish entirely by 2030. The oceans are becoming not just warmer but more acidic; the difference between daytime and nighttime temperatures is diminishing; animals are shifting their ranges poleward; and plants are blooming days, and in some cases weeks, earlier than they used to. These are the warning signs that the Charney panel cautioned against waiting for, and while in many parts of the globe they are still subtle enough to be overlooked, in others they can no longer be ignored. As it happens, the most dramatic changes are occurring in those places, like Shishmaref, where the fewest people tend to live. This disproportionate effect of global warming in the far north was also predicted by early climate models, which forecast, in column after column of FORTRAN-generated figures, what today can be measured and observed directly: the Arctic is melting.

Most of the land in the Arctic, and nearly a quarter of all the land in the Northern Hemisphere—some five and a half billion acres—is underlaid by zones of permafrost. A few months after I visited Shishmaref, I went back to Alaska to take a trip through the interior of the state with

Vladimir Romanovsky, a geophysicist and permafrost expert. I flew into Fairbanks—Romanovsky teaches at the University of Alaska, which has its main campus there—and when I arrived, the whole city was enveloped in a dense haze that looked like fog but smelled like burning rubber. People kept telling me that I was lucky I hadn't come a couple of weeks earlier, when it had been much worse. "Even the dogs were wearing masks," one woman I met said. I must have smiled. "I am not joking," she told me.

Fairbanks, Alaska's second-largest city, is surrounded on all sides by forest, and virtually every summer lightning sets off fires in these forests, which fill the air with smoke for a few days or, in bad years, weeks. In the summer of 2004, the fires started early, in June, and were still burning two and a half months later; by the time of my visit, in late August, a record 6.3 million acres—an area roughly the size of New Hampshire—had been incinerated. The severity of the fires was clearly linked to the weather, which had been exceptionally hot and dry; the average summertime temperature in Fairbanks was the highest on record, and the amount of rainfall was the third lowest.

On my second day in Fairbanks, Romanovsky picked me up at my hotel for an underground tour of the city. Like most permafrost experts, he is from Russia. (The Soviets more or less invented the study of permafrost when they decided to build their gulags in Siberia.) A broad man with shaggy brown hair and a square jaw, Romanovsky as a student had had to choose between playing professional hockey and becoming a geophysicist. He had opted for the latter, he told me, because "I was little bit better scientist than hockey player." He went on to earn two master's degrees and two Ph.D.s. Romanovsky came to get me at ten A.M.; owing to all the smoke, it looked like dawn.

Any piece of ground that has remained frozen for at least two years is, by definition, permafrost. In some places, like eastern Siberia, permafrost runs nearly a mile deep; in Alaska, it varies from a couple of hundred feet to a couple of thousand feet deep. Fairbanks, which is just below the Arctic Circle, is situated in a region of discontinuous permafrost, meaning that the city is pocked with regions of frozen ground. One of the first stops on Romanovsky's tour was a hole that had opened up in a patch of permafrost not far from his house. It was about six feet wide and five feet deep. Nearby were the outlines of other, even bigger holes, which, Romanovsky told me, had been filled with gravel by the local public-works department. The holes, known as thermokarsts, had appeared suddenly when the permafrost gave way, like a rotting floorboard. (The technical term for thawed permafrost is "talik," from a Russian word meaning "not frozen.") Across the road, Romanovsky pointed out a long trench running into the woods. The trench, he explained, had been formed when a wedge of underground ice had melted. The spruce trees that had been growing next to it, or perhaps on top of it, were now listing at odd angles, as if in a gale. Locally, such trees are called "drunken." A few of the spruces had fallen over. "These are very drunk," Romanovsky said.

In Alaska, the ground is riddled with ice wedges that were created during the last glaciation, when the cold earth cracked and the cracks filled with water. The wedges, which can be dozens or even hundreds of feet deep, tended to form in networks, so when they melt, they leave behind connecting diamond- or hexagon-shaped depressions. A few blocks beyond the drunken forest, we came to a house where the front yard showed clear signs of ice-wedge melt-off. The owner, trying to make the best of things, had turned the yard into a miniature-golf course. Around the corner, Romanovsky pointed out a house—no longer occupied—that basically had split in two; the main part was leaning to the right and the garage toward the left. The house had been built in the sixties or early seventies; it had survived until almost a decade ago, when the permafrost under it started to degrade. Romanovsky's mother-in-law used to own two houses on the same block. He had urged her to sell them both. He pointed out one, now under new ownership; its roof had developed an ominous-looking ripple. (When Romanovsky went to buy his own house, he looked only in permafrost-free areas.)

"Ten years ago, nobody cared about permafrost," he told me. "Now everybody wants to know." Measurements that Romanovsky and his colleagues at the University of Alaska have made around Fairbanks show that the temperature of the permafrost in many places has risen to the point where it is now less than one degree below freezing. In places where the permafrost has been disturbed, by roads or houses or lawns, much of it is already thawing. Romanovsky has also been monitoring the permafrost on the North Slope and has found that there, too, are regions where the permafrost is very nearly thirty-two degrees Fahrenheit. While thermokarsts in the roadbeds and talik under the basement are the sort of problems that really only affect the people right near—or above—them, warming permafrost is significant in ways that go far beyond local real estate losses. For one thing, permafrost represents a unique record of long-term temperature trends. For another, it acts, in effect, as a repository for greenhouse gases. As the climate warms, there is a good chance that these gases will be released into the atmosphere, further contributing to global warming. Although the age of permafrost is difficult to determine, Romanovsky estimates that most of it in Alaska probably dates back to the beginning of the last glacial cycle. This means that if it thaws, it will be doing so for the first time in more than a hundred and twenty thousand years. "It's really a very interesting time," Romanovsky told me.

The next morning, Romanovsky picked me up at seven. We were going to drive from Fairbanks nearly five hundred miles north to the town of Deadhorse, on Prudhoe Bay. Romanovsky makes the trip at least once a year, to collect data from the many electronic monitoring stations he has set up. Since the way was largely unpaved, he had rented a truck for the occasion. Its windshield was cracked in several places. When I suggested this could be a problem, Romanovsky assured me that it was "typical

Alaska." For provisions, he had brought along an oversize bag of Tostitos.

The road that we traveled along—the Dalton Highway—had been built for Alaskan oil, and the pipeline followed it, sometimes to the left, sometimes to the right. (Because of the permafrost, the pipeline runs mostly aboveground, on pilings that contain ammonia, which acts as a refrigerant). Trucks kept passing us, some with severed caribou heads strapped to their roofs, others belonging to the Alyeska Pipeline Service Company. The Alyeska trucks were painted with the disconcerting motto "Nobody Gets Hurt." About two hours outside Fairbanks, we started to pass through tracts of forest that had recently burned, then tracts that were still smoldering, and, finally, tracts that were still, intermittently, in flames. The scene was part Dante, part *Apocalypse Now*. We crawled along through the smoke. After another few hours, we reached Coldfoot, named, supposedly, for some gold prospectors who arrived at the spot in 1900, then got "cold feet" and turned around. We stopped to have lunch at a truck stop, which made up pretty much the entire town. Just beyond Coldfoot, we passed the tree line. An evergreen was marked with a plaque that read "Farthest North Spruce Tree on the Alaska Pipeline: Do Not Cut." Predictably, someone had taken a knife to it. A deep gouge around the trunk was bound with duct tape. "I think it will die," Romanovsky told me.

Finally, at around five P.M., we reached the turnoff for the first monitoring station. By now we were traveling along the edge of the Brooks Range and the mountains were purple in the afternoon light. Because one of Romanovsky's colleagues had nursed dreams—never realized—of traveling to the station by plane, it was situated near a small airstrip, on the far side of a quickly flowing river. We pulled on rubber boots and forded the river, which, owing to the lack of rain, was running low. The site consisted of a few posts sunk into the tundra, a solar panel, a two-hundred-foot-deep borehole with heavy-gauge wire sticking out of it, and a white container, resembling an ice chest, that held computer equipment. The solar panel, which the previous summer had been mounted a few feet off the ground, was now resting on the scrub. At first, Romanovsky speculated that this was a result of vandalism, but after inspecting things more closely, he decided that it was the work of a bear. While he hooked up a laptop computer to one of the monitors inside the white container, my job was to keep an eye out for wildlife.

For the same reason that it is sweaty in a coal mine—heat flux from the center of the earth—permafrost gets warmer the farther down you go. Under equilibrium conditions—which is to say, when the climate is stable—the very warmest temperatures in a borehole will be found at the bottom and temperatures will decrease steadily as you go higher. In these circumstances, the lowest temperature will be found at the permafrost's surface, so that, plotted on a graph, the results will be a tilted line. In recent decades, though, the temperature profile of

Alaska's permafrost has drooped. Now, instead of a straight line, what you get is shaped more like a sickle. The permafrost is still warmest at the very bottom, but instead of being coldest at the top, it is coldest somewhere in the middle, and warmer again toward the surface. This is a sign—and an unambiguous one—that the climate is heating up.

"It's very difficult to look at trends in air temperature, because it's so variable," Romanovsky explained after we were back in the truck, bouncing along toward Deadhorse. It turned out that he had brought the Tostitos to stave off not hunger but fatigue—the crunching, he said, kept him awake—and by now the enormous bag was more than half empty. "So one year you have around Fairbanks a mean annual temperature of zero"—thirty-two degrees Fahrenheit—"and you say, 'Oh yeah, it's warming,' and other years you have mean annual temperature of minus six"—twenty-one degrees Fahrenheit—"and everybody says, 'Where? Where is your global warming?' In the air temperature, the signal is very small compared to noise. What permafrost does is it works as low-pass filter. That's why we can see trends much easier in permafrost temperatures than we can see them in atmosphere." In most parts of Alaska, the permafrost has warmed by three degrees since the early 1980s. In some parts of the state, it has warmed by nearly six degrees.

When you walk around in the Arctic, you are stepping not on permafrost but on something called the "active layer."

The active layer, which can be anywhere from a few inches to a few feet deep, freezes in the winter but thaws over the summer, and it is what supports the growth of plants— large spruce trees in places where conditions are favorable enough and, where they aren't, shrubs and, finally, just lichen. Life in the active layer proceeds much as it does in more temperate regions, with one critical difference. Temperatures are so low that when trees and grasses die they do not fully decompose. New plants grow on top of the half-rotted old ones, and when these plants die the same thing happens all over again. Eventually, through a process known as cryoturbation, organic matter is pushed down beneath the active layer into the permafrost, where it can sit for thousands of years in a botanical version of suspended animation. (In Fairbanks, grass that is still green has been found in permafrost dating back to the middle of the last ice age.) This is the reason that permafrost, much like a peat bog or, for that matter, a coal deposit, acts as a storage unit for accumulated carbon.

One of the risks of rising temperatures is that the storage process can start to run in reverse. Under the right conditions, organic material that has been frozen for millennia will begin to break down, giving off carbon dioxide or methane, which is an even more powerful (though more short-lived) greenhouse gas. In parts of the Arctic, this process is already under way. Researchers in Sweden, for example, have been measuring the methane output of a bog known as the Stordalen mire, near the town of Abisko, nine hundred miles north of Stockholm, for almost thirty-five years. As the permafrost in the area has warmed, methane releases have increased, in some spots by as much as 60 percent. Thawing permafrost could make the active layer more hospitable to plants, which are a sink for carbon. Even this, though, wouldn't be enough to offset the release of greenhouse gases. No one knows exactly how much carbon is stored in the world's permafrost, but estimates run as high as 450 billion metric tons.

"It's like ready-use mix—just a little heat, and it will start cooking," Romanovsky told me. It was the day after we had arrived in Deadhorse, and we were driving through a steady drizzle out to another monitoring site. "I think it's just a time bomb, just waiting for a little warmer conditions." Romanovsky was wearing a rain suit over his canvas work clothes. I put on a rain suit that he had brought along for me. He pulled a tarp out of the back of the truck.

Whenever he has had funding, Romanovsky has added new monitoring sites to his network. There are now sixty of them, and while we were on the North Slope he spent all day and also part of the night—it stayed light until nearly eleven—rushing from one to the next. At each site, the routine was more or less the same. First, Romanovsky would hook up his computer to the data logger, which had been recording permafrost temperatures on an hourly basis since the previous summer. When it was raining, Romanovsky would perform this first step hunched under the tarp. Then he would take out a metal probe shaped like a "T" and poke it into the ground at regular intervals, measuring the depth of the active layer. The probe was a meter long, which, it turned out, was no longer quite long enough. The summer had been so warm that almost everywhere the active layer had grown deeper, in some spots by just a few centimeters, in other spots by more than that. In places where the active layer was particularly deep, Romanovsky had had to work out a new way of measuring it using the probe and a wooden ruler. (I helped out by recording the results of this exercise in his waterproof field notebook.) Eventually, he explained, the heat that had gone into increasing the depth of the active layer would work its way downward, bringing the permafrost that much closer to the thawing point. "Come back next year," he advised me.

On the last day I spent on the North Slope, a friend of Romanovsky's, Nicolai Panikov, a microbiologist at the Stevens Institute of Technology, in New Jersey, arrived. He was planning on collecting cold-loving microorganisms known as psychrophiles, which he would take back to New Jersey to study. Panikov's goal was to determine whether the organisms could have functioned in the sort of conditions that, it is believed, were once found on Mars. He told me that he was quite convinced that Martian life existed—or, at least, had existed. Romanovsky expressed his opinion on this by rolling his eyes; nevertheless, he had agreed to help Panikov dig up some permafrost.

That same day, I flew with Romanovsky by helicopter to a small island in the Arctic Ocean, where he had set up yet another monitoring site. The island, just north of the seventieth parallel, was a bleak expanse of mud dotted with little clumps of yellowing vegetation. It was filled with ice wedges that were starting to melt, creating a network of polygonal depressions. The weather was cold and wet, so while Romanovsky hunched under his tarp I stayed in the helicopter and chatted with the pilot. He had lived in Alaska since 1967. "It's definitely gotten warmer since I've been here," he told me. "I have really noticed that."

When Romanovsky emerged, we took a walk around the island. Apparently, in the spring it had been a nesting site for birds, because everywhere we went there were bits of eggshell and piles of droppings. The island was only about ten feet above sea level, and at the edges it dropped off sharply into the water. Romanovsky pointed out a spot along the shore where the previous summer a series of ice wedges had been exposed. They had since melted, and the ground behind them had given way in a cascade of black mud. In a few years, he said, he expected more ice wedges would be exposed, and then these would melt, causing further erosion. Although the process was different in its mechanics from what was going on in Shishmaref, it had much the same cause and, according to Romanovsky, was likely to have the same result. "Another disappearing island," he said, gesturing toward some freshly exposed bluffs. "It's moving very, very fast."

On September 18, 1997, the *Des Groseilliers*, a three-hundred-and-eighteen-foot-long icebreaker with a bright-red hull, set out from the town of Tuktoyaktuk, on the

Beaufort Sea, and headed north under overcast skies. Normally, the *Des Groseilliers*, which is based in Québec City, is used by the Canadian Coast Guard, but for this particular journey it was carrying a group of American geophysicists, who were planning to jam it into an ice floe. The scientists were hoping to conduct a series of experiments as they and the ship and the ice floe all drifted, as one, around the Arctic Ocean. The expedition had taken several years to prepare for, and during the planning phase its organizers had carefully consulted the findings of a previous Arctic expedition, which had taken place back in 1975. The researchers aboard the *Des Groseilliers* were aware that the Arctic sea ice was retreating; that was, in fact, precisely the phenomenon they were hoping to study. Still, they were caught off guard. Based on the data from the 1975 expedition, they had decided to look for a floe averaging nine feet thick. When they reached the area where they planned to overwinter—at seventy-five degrees north latitude—not only were there no floes nine feet thick, there were barely any that reached six feet. One of the scientists on board recalled the reaction on the *Des Groseilliers* this way: "It was like 'Here we are, all dressed up and nowhere to go.' We imagined calling the sponsors at the National Science Foundation and saying, 'Well, you know, we can't find any ice.'"

Sea ice in the Arctic comes in two varieties. There is seasonal ice, which forms in the winter and then melts in the summer, and perennial ice, which persists year-round. To the untrained eye, all of it looks pretty much the same, but by licking it you can get a good idea of how long a particular piece has been floating around. When ice begins to form in seawater, it forces out the salt, which has no place in the crystal structure. As the ice thickens, the rejected salt collects in tiny pockets of brine too highly concentrated to freeze. If you suck on a piece of first-year ice, it will taste salty. Eventually, if the ice stays frozen long enough, these pockets of brine drain out through fine, veinlike channels, and the ice becomes fresher. Multiyear ice is so fresh that if you melt it, you can drink it.

The most precise measurements of Arctic sea ice have been made by NASA, using satellites equipped with microwave sensors. In 1979, the satellite data show, perennial sea ice covered 1.7 billion acres, or an area nearly the size of the continental United States. The ice's extent varies from year to year, but since then the overall trend has been strongly downward. The losses have been particularly great in the Beaufort and Chukchi Seas, and also considerable in the Siberian and Laptev Seas. During this same period, an atmospheric circulation pattern known as the Arctic Oscillation has mostly been in what climatologists call a "positive" mode. The positive Arctic Oscillation is marked by low pressure over the Arctic Ocean, and it tends to produce strong winds and higher temperatures in the far north. No one really knows whether the recent behavior of the Arctic Oscillation is independent of global warming or a product of it. By now, though, the perennial sea ice has shrunk by roughly 250 million acres, an area the size of New York, Georgia, and Texas combined. According to mathematical models, even the extended period of a positive Arctic Oscillation can account for only part of this loss.

At the time the *Des Groseilliers* set off, little information on trends in sea-ice depth was available. A few years later, a limited amount of data on this topic—gathered, for rather different purposes, by nuclear submarines— was declassified. It showed that between the 1960s and the 1990s, sea-ice depth in a large section of the Arctic Ocean declined by nearly 40 percent.

Eventually, the researchers on board the *Des Groseilliers* decided that they would just have to settle for the best ice floe they could find. They picked one that stretched over some thirty square miles. In some spots it was six feet thick, in some spots just three. Tents were set up on the floe to house experiments, and a safety protocol was established: anyone venturing out onto the ice had to travel with a buddy and a radio. (Many also carried a gun, in case of polar-bear problems.) Some of the scientists speculated that, since the ice was abnormally thin, it would grow thicker during the expedition. Just the opposite turned out to be the case. The *Des Groseilliers* spent twelve months frozen into the floe, and, during that time, it drifted some three hundred miles north. Nevertheless, at the end of the year, the average thickness of the ice had declined, in some spots by as much as a third. By August 1998, so many of the scientists had fallen through that a new requirement was added to the protocol: anyone who set foot off the ship had to wear a life jacket.

*   *   *

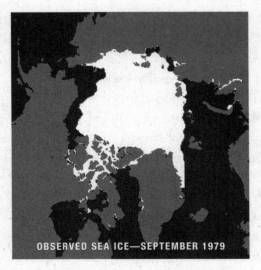

OBSERVED SEA ICE—SEPTEMBER 1979

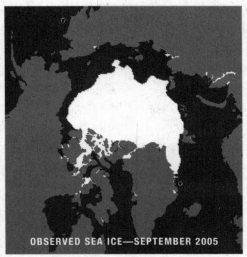

OBSERVED SEA ICE—SEPTEMBER 2005

The extent of the Arctic's perennial sea ice has declined dramatically in recent years. Credit: F. Fetterer and K. Knowles, Sea Ice Index, National Snow and Ice Data Center.

Donald Perovich has studied sea ice for thirty years, and on a rainy day not long after I got back from Deadhorse, I went to visit him at his office in Hanover, New Hampshire. Perovich works for the Cold Regions Research and Engineering Laboratory, or CRREL (pronounced "crell"). CRREL is a division of the U.S. Army that was established in 1961 in anticipation of a very cold war. (The assumption was that if the Soviets invaded, they would probably do so from the north.) He is a tall man with black hair, very black eyebrows, and an earnest manner. His office is decorated with photographs from the *Des Groseilliers* expedition, for which he served as the lead scientist; there are shots of the ship, the tents, and, if you look closely enough, the bears. One grainy-looking photo shows someone dressed up as Santa Claus, celebrating Christmas in the darkness out on the ice. "The most fun you could ever have" was how Perovich described the expedition to me.

Perovich's particular area of expertise, in the words of his CRREL biography, is "the interaction of solar radiation with sea ice." During the *Des Groseilliers* expedition, Perovich spent most of his time monitoring conditions on the floe using a device known as a spectroradiometer. Facing toward the sun, a spectroradiometer measures incident light, and facing toward earth, it measures reflected light. By dividing the latter by the former, you get a quantity known as albedo. (The term comes from the Latin word for "whiteness.") During April and May, when conditions on the floe were relatively stable, Perovich took measurements with his spectroradiometer once a week, and during June, July, and August, when they were changing more rapidly, he took measurements every other day. The arrangement allowed him to plot exactly how the albedo varied as the snow on top of the ice turned to slush, and then the slush became puddles, and, finally, some of the puddles melted through to the water below.

An ideal white surface, which reflected all the light that shone on it, would have an albedo of one, and an ideal black surface, which absorbed all the light, would have an albedo of zero. The albedo of the earth, in aggregate, is 0.3, meaning that a little less than a third of the sunlight that strikes it is reflected back out. Anything that changes the earth's albedo changes how much energy the planet absorbs, with potentially dramatic consequences. "I like it because it deals with simple concepts, but it's important," Perovich told me.

At one point, Perovich asked me to imagine that we were looking down at the earth from a spaceship hovering above the North Pole. "It's springtime, and the ice is covered with snow, and it's really bright and white," he said. "It reflects over 80 percent of the incident sunlight. The albedo's around 0.8, 0.9. Now, let's suppose that we melt that ice away and we're left with the ocean. The albedo of the ocean is less than 0.1; it's like 0.07.

"Not only is the albedo of the snow-covered ice high; it's the highest of anything we find on earth," he went on. "And not only is the albedo of water low; it's pretty much as low as anything you can find on earth. So what you're doing is you're replacing the best reflector with the worst reflector." The more open water that's exposed, the more solar energy goes into heating the ocean. The result is a positive feedback, similar to the one between thawing permafrost and carbon releases, only more direct. This so-called ice-albedo feedback is believed to be a major reason that the Arctic is warming so rapidly.

"As we melt that ice back, we can put more heat into the system, which means we can melt the ice back even more, which means we can put more heat into it, and, you see, it just kind of builds on itself," Perovich said. "It takes a small nudge to the climate system and amplifies it into a big change."

A few dozen miles to the east of CRREL, not far from the Maine–New Hampshire border, is a small park called the Madison Boulder Natural Area. The park's major—indeed, only—attraction is a block of granite the size of a two-story house. The Madison Boulder is thirty-seven feet wide and eighty-three feet long and weighs about ten million pounds. It was plucked out of the White Mountains and deposited in its current location eleven thousand years ago, and it illustrates how relatively minor changes to the climate system can, when amplified, yield monumental results.

Geologically speaking, we are now living in a warm period after an ice age. Over the past two million years, huge ice sheets have advanced across the Northern Hemisphere and retreated again more than twenty times. (Each major advance tended, for obvious reasons, to destroy the evidence of its predecessors.) The most recent advance, called the Wisconsin, began roughly 120,000 years ago. Ice began to creep outward from centers in Scandinavia, Siberia, and the highlands near Hudson Bay, spreading gradually across what is now Europe and Canada. By the time the sheets had reached their maximum southern extent, most of New England and New York and a good part of the upper Midwest were buried under ice nearly a mile thick. The ice sheets were so heavy that they depressed the crust of the earth, pushing it down into the mantle. (In some places, the process of recovery, called isostatic rebound, is still going on.) As the ice retreated, at the start of the current interglacial—the Holocene—it deposited, among other landmarks, the terminal moraine known as Long Island.

It is now known, or at least almost universally accepted, that glacial cycles are initiated by slight, periodic variations in the earth's orbit. These orbital variations, which are caused by, among other things, the gravitational pull of the other planets, alter the distribution of sunlight at different latitudes during different seasons and occur according to a complex cycle that takes a hundred thousand years to complete. Orbital variations in themselves, however, aren't sufficient to produce the sort of massive ice sheet that picked up the Madison Boulder.

The crushing size of that ice sheet, the Laurentide, which stretched over some five million square miles, was the result of feedbacks, more or less analogous to those now being studied in the Arctic, only operating in reverse. As the ice spread, albedo increased, leading to less heat absorption and the growth of yet more ice. At the same time, for reasons that are not entirely understood, as the ice sheets advanced, CO_2 levels declined: during each of the most recent glaciations, carbon dioxide levels dropped almost precisely in sync with falling temperatures. During each warm period, when the ice retreated, CO_2 levels rose again. Researchers who have studied this history have concluded that fully half the temperature difference between cold periods and warm ones can be attributed to changes in the concentrations of greenhouse gases.

While I was at CRREL, Perovich took me to meet a colleague of his named John Weatherly. Posted on Weatherly's office door was a bumper sticker designed to be pasted—illicitly—on SUVs. It said, I'M CHANGING THE CLIMATE! ASK ME HOW! Weatherly is a climate modeler, and for the past several years, he and Perovich have been working to translate the data gathered on the *Des Groseilliers* expedition into computer algorithms to be used in climate forecasting. Weatherly told me that some climate models—worldwide, there are about fifteen major ones in operation—predict that the perennial sea-ice cover in the Arctic will disappear entirely by the year 2080. At that point, although there would continue to be seasonal ice that forms in winter, in summer the Arctic Ocean would be completely ice-free. "That's not in our lifetime," he observed. "But it is in the lifetime of our kids."

Later, back in his office, Perovich and I talked about the

long-term prospects for the Arctic. Perovich noted that the earth's climate system is so vast that it is not easily altered. "On the one hand, you think, It's the earth's climate system; it's big, it's robust. And, indeed, it has to be somewhat robust or else it would be changing all the time." On the other hand, the climate record shows that it would be a mistake to assume that change, when it comes, will come gradually. Perovich offered a comparison that he had heard from a glaciologist friend. The friend likened the climate system to a rowboat: "You can tip and then you'll just go back. You can tip it and just go back. And then you tip it and you get to the other stable state, which is upside down."

Perovich said that he also liked a regional analogy. "The way I've been thinking about it, riding my bike around here, is, You ride by all these pastures and they've got these big granite boulders in the middle of them. You've got a big boulder sitting there on this rolling hill. You can't just go by this boulder. You've got to try to push it. So you start rocking it, and you get a bunch of friends, and they start rocking it, and finally it starts moving. And then you realize, Maybe this wasn't the best idea. That's what we're doing as a society. This climate, if it starts rolling, we don't really know where it will stop."

and a few months later was elected to a professorship in natural philosophy. His lectures were enormously popular—many were collected and published—a fact that testifies both to Tyndall's considerable skills as a speaker and also to the intellectual interests of the Victorian middle class. Eventually, Tyndall went on a lucrative speaking tour of the United States, the proceeds from which he placed in a special trust to be used for the advancement of American science.

Tyndall's research varied almost impossibly widely, from optics to acoustics to glacial motion. (He was an avid mountain climber, and made frequent trips to the Alps to study the ice.) One of his most enduring interests was in the science of heat, which, in the mid-nineteenth century, was rapidly evolving. In 1859, Tyndall built the world's first ratio spectrophotometer, a device that allowed him to compare the way different gases absorb and transmit radiation. When Tyndall tested the most common gases in the air—nitrogen and oxygen—he found they were transparent to both visible and infrared radiation. (The latter of these he called "ultra-red" radiation.) Other gases, like carbon dioxide, methane, and water vapor, however, were not. CO_2 and water vapor were transparent in the visible part of the spectrum, but partly opaque in the infrared. Tyndall was quick to appreciate the implications of his discovery: the selectively transparent gases, he declared, were largely responsible for determining the planet's climate. He likened their impact to that of a dam built across a river: just as a dam "causes a local deepening of the stream, so our atmosphere, thrown as a barrier across the terrestrial rays, produces a local heightening of the temperature at the earth's surface."

The phenomenon that Tyndall identified is now referred to as the "natural greenhouse effect." It is not remotely controversial; indeed, it's recognized as an essential condition of life on the planet. To understand how it works, it helps to imagine the world without it. In that situation, the earth would be constantly receiving energy from the sun and, at the same time, constantly radiating energy back out to space. All hot bodies radiate, and the amount that they radiate is a function of their temperature. (The exact relationship is expressed by a formula known as

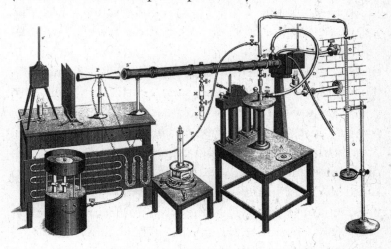

The world's first ratio spectrophotometer, built by John Tyndall, was used to measure the absorptive properties of gases.
Credit: Philosophical Transactions, *vol. 151 (1861).*

the Stefan-Boltzmann law, which states that the radiation emitted by an object is proportional to its absolute temperature raised to the fourth power: $P/A = \sigma T^4$*.) In order for the earth to be in equilibrium, the quantity of energy it radiates out into space must equal the quantity of radiation it is receiving. When, for whatever reason, equilibrium is disturbed, the planet will either warm up or cool down until its temperature is once again sufficient to make the two energy streams balance out.

If there were no greenhouse gases in the atmosphere, energy radiating from the surface of the earth would flow away unimpeded. In that case, it would be comparatively easy to calculate how warm the planet would have to be to throw back into space the same amount of energy it receives from the sun. (This amount varies widely by location and time of year; averaged out over all latitudes and all seasons it comes to some 235 watts per square meter, or roughly the power of four household lightbulbs.) The result of this calculation turns out to be a frigid zero degrees. To use Tyndall's Victorian language, if the heat-trapping gases were removed from the earth's atmosphere, "the warmth of our fields and gardens would pour itself unrequited into space, and the sun would rise upon an island held fast in the iron grip of frost."

Greenhouse gases alter the situation because of their selectively absorptive properties. They allow the sun's radiation, which arrives mostly in the form of visible light,

* P stands for power, in watts; A for area, in square meters; T for temperature, in degrees Kelvin. σ is the Stefan-Boltzmann constant, 5.67×10^{-8} W/m^{2}K^4.

to pass freely. But the earth's radiation, which is emitted in the infrared part of the spectrum, is partially blocked. Greenhouse gases absorb infrared radiation and then re-emit it—some out toward space and some back toward earth. This process of absorption and re-emission has the effect of limiting the outward flow of energy; as a result, the earth's surface and its lower atmosphere need to be that much warmer before the planet can radiate out the necessary 235 watts per square meter. The presence of greenhouse gases largely accounts for the fact that the average global temperature, instead of zero, is actually a far more comfortable fifty-seven degrees.

Tyndall suffered from insomnia, which grew worse as he grew older, and in 1893 he died from an overdose of chloral hydrate—an early sleeping drug—that had been administered by his wife. ("My poor darling, you have killed your John," he is reported to have told her shortly before expiring.) Right around the time of his poisoning, the Swedish chemist Svante Arrhenius took up where he had left off.

Arrhenius would eventually come to be regarded as one of the giants of nineteenth-century science, but his career, like Tyndall's, began inauspiciously. In 1884, when Arrhenius was a student at the University of Uppsala, he wrote a doctoral dissertation on the behavior of electrolytes. (In 1903, he would be awarded the Nobel Prize for this work, now known as the theory of electrolytic dissociation.) The university's examining committee was so unimpressed that it awarded the dissertation a fourth-class mark: *non sine laude*. Arrhenius spent the next several years bouncing from one foreign post to another before finally being offered a teaching position back home in Sweden. He would not be elected to the Swedish Academy of Sciences until shortly before winning the Nobel Prize, and even then his election faced strong opposition.

Why, exactly, Arrhenius became curious about the effects of CO_2 on global temperatures is unclear; mainly he seems to have been interested in determining whether falling carbon dioxide levels could have caused the ice ages. (Some biographers have noted, although it's hard to find any real connection, that his work on the subject coincided with his separation from his wife—earlier his student—who had taken their only son with her.) Tyndall had recognized the influence of greenhouse gas levels on the climate, and indeed had even pro- posed—presciently, but not entirely correctly—that var- iations in these levels would have been capable of producing "all the mutations of climate which the researches of geologists reveal." But Tyndall never went beyond such qualitative speculations. Arrhenius decided to actually calculate how the earth's temperature would be affected by changing CO_2 levels. He would later describe this task as one of the most tedious of his life. He began working on it on Christmas Eve 1894, and although he routinely toiled for fourteen hours a day— "I have not worked this hard since I was cramming for my B.A.," he wrote to a friend—he was not finished for nearly a year. Finally, in December 1895, he was ready to present his conclusions to the Swedish Academy.

By today's standards, Arrhenius's work seems primitive. All of his calculations were performed using pen and paper. He was missing crucial pieces of information about spectral absorption, and he ignored several potentially important feedbacks. These deficiencies, however, seem more or less to have canceled each other out. Arrhenius asked what would happen to the earth's climate if CO_2 levels were halved and also if they were doubled. In the case of doubling, he determined that average global temperatures would rise between nine and eleven degrees, a result that approximates the estimates of the most sophisticated climate models in operation today.

Arrhenius was also responsible for a key conceptual breakthrough. All over Europe, factories and railroads and power stations were burning coal and belching out smoke. Arrhenius recognized that industrialization and climate change were intimately related, and that the consumption of fossil fuels must, over time, lead to warming. He was not, however, terribly concerned about this. Arrhenius thought that the buildup of carbon dioxide in the air would be extremely slow—at one point, he estimated that it would take three thousand years of coal burning to double atmospheric levels—mostly because he believed the oceans would act as a vast sponge, soaking up extra CO_2. Perhaps owing to the age he lived in, or perhaps just because he was Scandinavian, he anticipated that the results would, on the whole, be salubrious. Addressing the Swedish Academy,

Arrhenius declared that rising levels of carbon dioxide, which at the time was referred to as "carbonic acid," would allow future generations "to live under a warmer sky." Later, he elaborated on this notion in one of his numerous works of popular science, *Worlds in the Making*:

> By the influence of the increasing percentage of carbonic acid in the atmosphere we may hope to enjoy ages with more equable and better climates, especially as regards the colder regions of the earth, ages when the earth will bring forth much more abundant crops than at present for the benefit of rapidly propagating mankind.

After Arrhenius's death, in 1927, interest in climate change dropped off. Most scientists continued to believe that if carbon dioxide levels were rising at all, they were rising very slowly. Then, in the mid-1950s, for no particularly good reason, a young chemist named Charles David Keeling decided to work out a new and more precise way of measuring atmospheric CO_2. (Later he would explain his decision by saying he was "having fun" trying to assemble the necessary equipment.) In 1958, Keeling convinced the U.S. Weather Bureau to start using his technique to monitor CO_2 at its new observatory, eleven thousand feet above sea level, on the flank of Mauna Loa, on the island of Hawaii. These same CO_2 measurements have been taken at Mauna Loa nearly continuously ever since. The results, known as the "Keeling Curve," may well be the most widely reprinted set of natural science data ever collected.

Presented in the form of a graph, the Keeling Curve looks like the edge of a saw that is being held at a tilt. Each tooth on the saw corresponds to a single year. CO_2 levels fall to a minimum in the summer, when the trees of the Northern Hemisphere are taking up carbon dioxide for photosynthesis, and rise to a maximum in the winter, when these trees go dormant. (In the Southern Hemisphere, there are fewer forests.) The tilt, meanwhile, corresponds to the rising annual mean.

The first full year that CO_2 levels were recorded at Mauna Loa—1959—that mean stood at 316 parts per million. By the following year, it had reached 317 parts per million, prompting Keeling to observe that the reigning assumption about CO_2 absorption by the oceans was

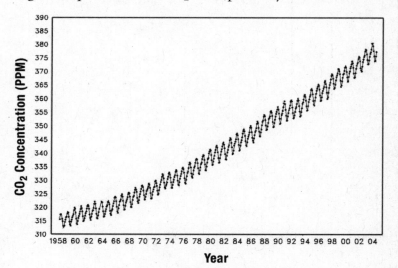

The Keeling Curve shows that CO_2 levels have been rising steadily since the 1950s. Credit: Scripps Institution of Oceanography.

probably wrong. By 1970, the level had reached 325 parts per million, and by 1990, it was up to 354 parts per million. In the summer of 2005, the CO_2 level stood at 378 parts per million, and by now, it has almost certainly risen to 380 parts per million. At this rate, it will reach 500 parts per million—nearly double preindustrial levels—by the middle of this century, which is to say, roughly two thousand eight hundred and fifty years ahead of Arrhenius's prediction.

Chapter 3

UNDER THE GLACIER

S WISS CAMP IS a research station that was set up in 1990 on a platform drilled into the Greenland ice sheet. Ice flows like water, only more slowly, and, as a result, the camp is always in motion: in fifteen years, it has migrated by more than a mile, generally in a westerly direction. Every summer, the whole place gets flooded, and every winter, its contents solidify. The cumulative effect of all this is that almost nothing at Swiss Camp functions anymore the way it was supposed to. To get into it, you have to clamber up a snowdrift and descend through a trapdoor in the roof, as if entering a ship's hold or a space module. The living quarters are no longer habitable, so now everyone at the camp sleeps outside, in tents. (The one assigned to me was, I was told, the same sort used by Robert Scott on his ill-fated expedition to the South Pole.) By the time I arrived at the camp, in late May, someone had jackhammered out the center of the workspace, which was equipped with some battered conference tables. Under the tables, where, under normal circumstances, you would stick your legs, there were still three-foot-high blocks of ice. Inside of the blocks, I could dimly make out a tangle of wires, a bulging plastic bag, and an old dustpan.

Konrad Steffen, a professor of geography at the University of Colorado, is the director of Swiss Camp. A native of Zurich, Steffen speaks English in the lilting cadences of *Schweizerdeutsch*. He is tall and lanky, with pale blue eyes, a graying beard, and the unflappable manner of a cowboy in a western. Steffen fell in love with the Arctic when, as a graduate student in 1975, he spent a summer on Axel Heiberg Island, four hundred miles northwest of the north magnetic pole. A few years later, for his doctoral dissertation, he lived for two winters on the sea ice off Baffin Island. (Steffen told me that for his honeymoon he had wanted to take his wife to Spitsbergen, an island five hundred miles north of Norway, but she demurred, and they ended up driving across the Sahara instead.)

When Steffen planned Swiss Camp—he built much of the place himself—it was not with global warming in mind. Rather, he was interested in following meteorological conditions on what is known as the ice sheet's "equilibrium line." Along this line, winter snow and summer melt are supposed to be precisely in balance. But in recent years, "equilibrium" has become an increasingly elusive quality. During the summer of 2002, for example, melt occurred in areas where liquid water had not been seen for hundreds, perhaps thousands, of years. The following winter, there was an unusually low snowfall, and in the summer of 2003, the melt was so great that, around Swiss Camp, five feet of ice were lost.

When I arrived at the camp, the 2004 melt season was already under way. This, to Steffen, was a matter of both intense scientific interest and serious practical concern. A few days earlier, one of his graduate students, Russell Huff, and a postdoc, Nicolas Cullen, had driven out on snow-mobiles to service some weather stations closer to the coast. The snow there was warming so fast that they had had to work until five in the morning, and then take a long detour back, to avoid getting caught in the quickly forming rivers. Steffen wanted to complete everything that needed to be done ahead of schedule, in case everyone had to pack up and leave early. My first day at Swiss Camp he spent fixing an antenna that had fallen over in the previous year's melt. It was bristling with equipment, like a high-tech Christmas tree. Even on a relatively mild day on the ice sheet, which this was, it never gets more than a few degrees above freezing, and I was walking around in a huge parka, two pairs of pants plus long underwear, and two pairs of gloves. Steffen, meanwhile, was tinkering with the antenna with his bare hands. He had spent the last fourteen summers at Swiss Camp, and I asked him what he had learned during that time. He answered with another question.

"Are we disintegrating part of the Greenland ice sheet over the longer term?" he asked. He was sorting through a tangle of wires that to me all looked the same but must have had some sort of distinguishing characteristics. "What the regional models tell us is that we will get more melt at the coast. It will continue to melt. But warmer air can hold more water vapor, and at the top of the ice sheet you'll get more precipitation. So we'll add more snow there. We'll get an imbalance of having more accumulation at the top, and more melt at the bottom. The key question now is: What is the dominant one, the more melt or the increase?"

Greenland, the world's largest island, is nearly four times the size of France—840,000 square miles—and, except for its southern tip, lies entirely above the Arctic Circle. The first Europeans to make a stab at settling it were the Norse, under the leadership of Erik the Red, who, perhaps deliberately, gave the island its misleading name. In the year 985, he arrived with twenty-five ships and nearly seven hundred followers. (Erik had left Norway when his father was exiled for killing a man, and then was himself exiled from Iceland for killing several more.) The Norse established two settlements: the Eastern Settlement, which was actually in the south, and the Western Settlement, which was to the north. For roughly four hundred years, they managed to scrape by, hunting, raising livestock, and making occasional logging expeditions to the coast of Canada. But then something went wrong. The last written record of them is an Icelandic affidavit regarding the marriage of Thorstein Ólafsson and Sigridur Björnsdóttir, which took place in the Eastern Settlement on the "Second Sunday after the Mass of the Cross," in the autumn of 1408.

These days the island has just over fifty-six thousand inhabitants, most of them Inuit, and almost a quarter live in the capital, Nuuk, about four hundred miles up the western coast. Since the late 1970s, Greenland has enjoyed a measure of home rule, but the Danes, who consider the island a province, still spend more than three hundred million dollars a year to support it. The result is a thin and not entirely convincing first-world veneer. Greenland has almost no agriculture, or industry, or, for that matter, roads. Following Inuit tradition, private ownership of land is not allowed, although it is possible to buy a house, an expensive proposition in a place where even the sewage pipes have to be insulated.

More than 80 percent of Greenland is covered by ice. Locked into this enormous glacier is 8 percent of the world's fresh water supply. Except for researchers like those at Swiss Camp, no one lives on the ice, or even ventures out onto it very often. (The edges are riddled with crevasses large enough to swallow a dogsled or, should the occasion arise, a five-ton truck.)

Like all glaciers, the Greenland ice sheet is made up entirely of accumulated snow. The most recent layers are thick and airy, while the older layers are thin and dense, which means that to drill through the ice is to descend backward in time, at first gradually, and then much more rapidly. A hundred and thirty-eight feet down, there is snow that fell during the time of the American Civil War; 2,500 feet down, snow from the time of the Peloponnesian Wars, and, 5,350 feet down, snow from the days when the cave painters of Lascaux were slaughtering bison. At the very bottom, 10,000 feet down, there is snow that fell on central Greenland before the start of the last ice age, more than a hundred thousand years ago.

As the snow is compressed, its crystal structure changes to ice. (Two thousand feet down, there is so much pressure on the ice that a sample drawn to the surface will, if mishandled, fracture, and in some cases even explode.) But in most other respects, the snow remains unchanged, a relic of the climate that first formed it. In the Greenland ice, there is nuclear fallout from early atomic tests, volcanic ash from Krakatau, lead pollution from ancient Roman smelters, and dust blown in from Mongolia on ice age winds. Every layer also contains tiny bubbles of trapped air, each of them a sample of a past atmosphere.

Much of what is known about the earth's climate over the last hundred thousand years comes from ice cores drilled in central Greenland, along a line known as the ice divide. Owing to differences between summer and winter snow, each layer in a Greenland core can be individually dated, much like the rings of a tree. Then, by analyzing the isotopic composition of the ice, it is possible to determine how cold it was at the time each layer was formed. Over the last decade, three Greenland cores have been drilled to a depth of nearly two miles, and these cores have prompted a wholesale rethinking of how the climate operates. Where once the system was thought to change, as it were, only glacially, now it is known to be capable of sudden and unpredictable reversals. One such reversal, called the Younger Dryas, after a small Arctic plant—*Dryas octopetala*—that suddenly reappeared in Scandinavia, took place roughly 12,800 years ago.

At that point, the earth, which had been warming rapidly, was plunged back into ice age conditions. It remained frigid for twelve centuries and then warmed again, even more abruptly. In Greenland, average annual temperatures shot up by nearly twenty degrees in a single decade.

As a continuous temperature record, the Greenland ice cores stop providing reliable information right around the start of the last glaciation. Climate records pieced together from other sources indicate that the previous interglacial,

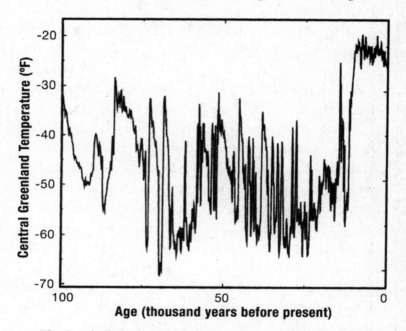

The Greenland record reveals that temperatures have often swung wildly. Credit: The Two-Mile Time Machine, *Princeton University Press, after K. Cuffey and G. Clow,* Journal of Geophysical Research, *vol. 102 (1997).*

which is known as the Eemian, was somewhat warmer than the present one, the Holocene. They also show that sea levels during that time were at least fifteen feet higher than they are today. One theory attributes this to a collapse of the West Antarctic ice sheet. A second holds that meltwater from Greenland was responsible. (When sea ice melts, it does not affect sea level, because the ice, which was floating, was already displacing an equivalent volume of water.) All told, the Greenland ice sheet holds enough water to raise sea levels worldwide by twenty-three feet. Scientists at NASA have calculated that throughout the 1990s the ice sheet, despite some thickening at the center, was shrinking by twelve cubic miles per year.

Jay Zwally is a NASA scientist who works on a satellite project known as the Ice Cloud and Land Elevation Satellite (ICESat). He is short and stocky, with a round face and a mischievous grin. Zwally is a friend of Steffen's and about ten years ago, he got the idea of installing global-positioning-system receivers around Swiss Camp to study changes in the ice sheet's elevation. He happened to be at the camp at the same time I was, and the second day of my visit we all got onto snowmobiles and headed out to a location known as JAR 1 (for Jakobshavn Ablation Region) to reinstall a GPS receiver. The trip was about ten miles. Midway through it, Zwally told me that he had once seen spy-satellite photos of the region we were crossing, and that they had shown that underneath the snow it was full of crevasses. Later, when I asked Steffen about this, he told me that he had had the whole area surveyed with bottom-seeking radar, and no crevasses of any note had been found. I was never sure which one of them to believe.

Reinstalling Zwally's GPS receiver entailed putting up a series of poles, a process that, in turn, required drilling holes thirty feet down into the ice. The drilling was done not mechanically but thermally, using a steam drill that consisted of a propane burner, a steel tank, and a long rubber hose. Everyone—Steffen, Zwally, the graduate students, me—took a turn. This meant holding on to the hose while it melted its way down, an activity reminiscent of ice fishing. Seventy-five years ago, not far from JAR 1, Alfred Wegener, the German scientist who proposed the theory of continental drift, died while on a meteorological expedition. He was buried in the ice sheet, and there is a running joke at Swiss Camp about stumbling onto his body. "It's Wegener!" one of the graduate students exclaimed, as the drill worked its way downward. The first hole was finished relatively quickly, at which point everyone decided—prematurely, as it turned out—that it was time for a midday break. Unless a hole stays filled with water, it starts to close up again, and can't be used. Apparently, there were fissures in the ice, because water kept draining out of the next few holes that were tried. The original plan had been for three holes, but, some six hours later, only two had been drilled, and it was decided that this would have to suffice.

Although Zwally had set out to look for changes in the

ice sheet's elevation, what he ended up discovering was even more significant. His GPS data showed that as the ice sheet melted, it didn't so much sink as start to accelerate. Thus, in the summer of 1996, the ice around Swiss Camp moved at a rate of thirteen inches per day, but, in 2001, it had sped up to twenty inches per day. The reason for this acceleration, it is believed, is that meltwater from the surface makes its way down to the bedrock below, where it acts as a lubricant. (In the process, it enlarges cracks and forms huge ice tunnels, known as "moulins.") Zwally's measurements also showed that, in the summer, the ice sheet rises by about six inches, indicating that it is floating on a cushion of water.

At the end of the last glaciation, the ice sheets that covered much of the Northern Hemisphere disappeared in a matter of a few thousand years—a surprisingly short time, considering how long it had taken them to build up. At one point, about fourteen thousand years ago, they were melting so fast that sea levels were rising at the rate of more than a foot a decade. Just how this happened is not entirely understood, but the acceleration of the Greenland ice sheet suggests yet another feedback mechanism: once an ice sheet begins to melt, it starts to flow faster, which means it also thins out faster, encouraging further melt. Not far from Swiss Camp is the huge river of ice known as the Jakobshavn Isbrae. In 1992, the Jakobshavn Isbrae flowed at a rate of 3.5 miles per year; by 2003, its velocity had increased to 7.8 miles per year. (Similar findings were announced recently by scientists measuring the flow of ice streams on the Antarctic Peninsula.) On the basis of Zwally's findings, James Hansen, the NASA official who directed one of the initial 1970s studies on the effects of carbon dioxide, has argued that if greenhouse gas emissions are not controlled, the total disintegration of the Greenland ice sheet could be set in motion in a matter of decades. Although the process could take centuries to fully play out, once begun it would become self-reinforcing, and hence virtually impossible to stop. In an article published in the journal *Climatic Change* in February 2005, Hansen, who is now the head of the Goddard Institute for Space Studies, wrote that he hoped he was wrong about the ice sheet, but added, "I doubt it."

As it happened, I was at Swiss Camp just as the global-warming disaster movie *The Day After Tomorrow* was opening in theaters. One night, Steffen's wife called on the camp's satellite phone to say that she had just taken the couple's two teenage children to see it. Everyone had enjoyed the film, she reported, especially because of the family connection.

The fantastic conceit of *The Day After Tomorrow* is that global warming produces global freezing. At the start of the film, a chunk of Antarctic ice the size of Rhode Island suddenly melts. (Something very similar to this actually happened in March 2002, when the Larsen B ice shelf collapsed.) Most of what follows—an instant ice age, cyclonic winds that descend from the upper atmosphere—is impossible as science but not as metaphor.

The record preserved in the Greenland ice sheet shows that our own relatively static experience of climate is actually what is exceptional. During the last glaciation, even as much of the world was frozen solid, average temperatures in Greenland frequently shot up, or down, by ten degrees, as in the Younger Dryas. Nobody knows what caused the sudden climate shifts of the past; however, many climatologists suspect that they had something to do with changes in ocean-current patterns that are known as the "thermohaline circulation."

"When you freeze sea ice, the salt is pushed out of the pores, so that the salty water actually drains," Steffen explained to me one day when we were standing out on the ice, not far from camp, trying to talk above the howl of the wind. "And salty water's actually heavier, so it starts to sink." Meanwhile, owing both to evaporation and cooling, water from the tropics becomes denser as it drifts toward the Arctic; near Greenland a tremendous volume of seawater is constantly sinking toward the ocean floor. As a result of this process, still more warm water is drawn from the tropics toward the poles, setting up what is often referred to as a "conveyor belt" that moves vast amounts of heat around the globe.

"This is the energy engine for the world climate," Steffen went on. "And it has one source: the water that sinks down. And if you just turn the knob here a little bit"—he made a motion of turning the water on in a bathtub—"we can expect significant temperature changes based on the redistribution of energy." One way to turn the knob is to heat the oceans, which is already happening. Another is to pour more freshwater into the polar seas. This is also occurring. Not only is runoff from coastal Greenland increasing; the volume of river discharge into the Arctic Ocean has been rising. Oceanographers monitoring the North Atlantic have documented that in recent decades its waters have become significantly less salty. A total shutdown of the thermohaline circulation is considered extremely unlikely in the coming century. But, if the Greenland ice sheet were to start to disintegrate, the possibility of such a shutdown could not be ruled out. Wallace Broecker, a professor of geochemistry at Columbia University's Lamont-Doherty Earth Observatory, has labeled the thermohaline circulation the "Achilles' heel of the climate system." Were it to halt, places like Britain, whose climate is heavily influenced by the Gulf Stream, could become much colder, even as the planet as a whole continued to warm up.

For the whole time I was at Swiss Camp, it was "polar day," and so the sun never set. Dinner was generally served at ten or eleven P.M., and afterward everyone sat around a makeshift table in the kitchen, talking and drinking coffee. (Because it weighs a lot and is not—strictly speaking—necessary, alcohol was in short supply.) One night, I asked Steffen what he thought conditions at Swiss Camp would be like in the same season a decade hence. "In ten years, the signal should be much more distinct, because we will have added another ten years of greenhouse warming," he said.

Zwally interjected, "I predict that ten years from now we won't be coming this time of year. We won't be able to come this late. To put it nicely, we are heading into deep doo-doo."

Either by disposition or by training, Steffen was reluctant to make specific predictions, whether about Greenland or, more generally, about the Arctic. Often, he prefaced his remarks by noting that there could be a change in atmospheric-circulation patterns that would dampen the rate of temperature increase or even—temporarily, at least—reverse it entirely. But he was emphatic that "climate change is a real thing."

"It's not something dramatic now—that's why people don't really react," he told me. "But if you can convey the message that it will be dramatic for our children and our children's children—the risk is too big not to care." The time, he added, "is already five past midnight."

On the last night that I spent at Swiss Camp, Steffen took the data he had downloaded off his weather station and ran them through various programs on his laptop to produce the mean temperature at the camp for the previous year. It was, it turned out, the highest of any year since the camp was built. When Steffen announced this to the group around the kitchen table, no one seemed the slightest bit surprised.

That night, dinner was unusually late. On the return trip of another pole-drilling expedition, one of the snowmobiles had caught on fire, and had had to be towed back to camp. When I finally went out to my tent to go to bed, I found that the snow underneath it had started to melt, and there was a large puddle in the middle of the floor. I went back to the kitchen to get some paper towels and tried to mop it up. But the puddle was too big, and eventually I gave up.

No nation takes a keener interest in climate change, at least on a per-capita basis, than Iceland. More than 10 percent of the country is covered by glaciers, the largest of which, Vatnajökull, stretches over thirty-two hundred square miles. During the so-called Little Ice Age, which began in Europe some five hundred years ago and ended some three hundred and fifty years later, the advance of the glaciers caused widespread misery. Contemporary records tell of farms being buried under the ice—"Frost and cold torment people," a pastor in eastern Iceland named Olafur Einarsson wrote—and in particularly severe years, shipping, too, seems to have ceased, because the island remained icebound even in summer. In the mid-eighteenth century, it has been estimated, nearly a third of the country's population died of starvation or associated cold-related ills. For Icelanders, many of whom can trace their genealogy back a thousand years, this is considered to be almost recent history.

Oddur Sigurdsson heads up a group called the Icelandic Glaciological Society. On a dark and dreary autumn afternoon, I went to visit him in his office, at the headquarters of Iceland's National Energy Authority, in Reykjavík. Little towheaded children kept wandering in to peer under

his desk, and then wandering out again, giggling. Sigurdsson explained that Reykjavík's public school teachers were
on strike, and his colleagues had had to bring their children
to work with them.

The Icelandic Glaciological Society is composed entirely of volunteers. Every fall, after the summer-melt
season has ended, they survey the size of the country's
three-hundred-odd glaciers and then file reports, which
Sigurdsson collects in brightly colored binders. In the
organization's early years—it was founded in 1930—the
volunteers were mostly farmers; they took measurements
by building cairns and pacing off the distance to the
glacier's edge. These days, members come from all walks
of life—one is a retired plastic surgeon—and they take
more exacting surveys, using tape measures and iron poles.
Some glaciers have been in the same family, so to speak, for
generations. Sigurdsson became head of the society in
1987, at which point one volunteer told him that he
thought he would like to relinquish his post.

"He was about ninety when I realized how old he was,"
Sigurdsson recalled. "His father had done this at that place
before and then his nephew took over for him." Another
volunteer has been monitoring his glacier, a section of
Vatnajökull, since 1948. "He's eighty," Sigurdsson said.
"And if I have some questions that go beyond his age, I just
go and ask his mother. She's a hundred and seven."

In contrast to glaciers in North America, which have
been shrinking steadily since the 1960s, Iceland's glaciers
grew through the 1970s and '80s. Then, in the mid-1990s,

they, too, began to contract. Sigurdsson pulled out a notebook of glaciological reports, filled out on yellow forms, and turned to the section on a glacier called Sólheimajökull, a tongue-shaped spit of ice that sticks out from a much larger glacier known as Mýrdalsjökull. In 1996, Sólheimajökull crept back by 10 feet. In 1997, it receded by another 33 feet, and in 1998 by 98 feet. Every year since then, it has retreated even more. In 2003, it shrank by 302 feet, and in 2004, by 285 feet. All told, Sólheimajökull—the name means "sun-home glacier" and refers to a nearby farm—is now 1,100 feet shorter than it was just a decade ago. Sigurdsson pulled out another notebook, which was filled with slides. He picked out some recent ones of Sólheimajökull. The glacier ended in a wide river. An enormous rock, which Sólheimajökull had deposited when it began its retreat, stuck out from the water like the hull of an abandoned ship.

"You can tell by this glacier what the climate is doing," Sigurdsson said. "It is more sensitive than the most sensitive meteorological measurement." He introduced me to a colleague of his, Kristjana Eythórsdóttir, who, as it turned out, was the granddaughter of the founder of the Icelandic Glaciological Society. Eythórsdóttir keeps tabs on a glacier named Leidarjökull, which is a four-hour trek from the nearest road. I asked her how it was doing. "Oh, it's getting smaller and smaller, just like all the others," she said. Sigurdsson told me that climate models predicted that by the end of the next century Iceland would be virtually ice-free. "We will have small ice caps on the highest mountains, but the mass of the glaciers will have gone," he said. It is believed that there have been glaciers on Iceland for at least the last two million years. "Probably longer," Sigurdsson said.

In October 2000, in a middle school in Barrow, Alaska, officials from the eight Arctic nations—the United States, Russia, Canada, Denmark, Norway, Sweden, Finland, and Iceland—met to talk about global warming. The group announced plans for a three-part, two-million-dollar study of climate change in the region. In November 2004, the first two parts of the study—a massive technical document and a hundred-and-forty-page summary—were presented at a symposium in Reykjavík.

The day after I went to talk to Sigurdsson, I attended the symposium's plenary session. In addition to nearly three hundred scientists, it drew a sizable contingent of native Arctic residents—reindeer herders, subsistence hunters, and representatives of groups like the Inuvialuit Game Council. In among the shirts and ties, I spotted two men dressed in the brightly colored tunics of the Sami and several others wearing sealskin vests. As the session went on, the subject kept changing—from hydrology and bio-diversity to fisheries and on to forests. The message, however, stayed the same. Almost wherever you looked, conditions in the Arctic were changing, and at a rate that surprised even those who had expected to find clear signs of warming. Robert Corell, an American oceanographer and former assistant director at the National Science

Foundation, coordinated the study. In his opening remarks, he ran through its findings—shrinking sea ice, receding glaciers, thawing permafrost—and summed them up as follows: "The Arctic climate is warming rapidly now, with an emphasis on *now*." Particularly alarming, Corell said, were the most recent data from Greenland, which showed the ice sheet melting much faster "than we thought possible even a decade ago."

Global warming is routinely described as a matter of scientific debate—a theory whose validity has yet to be demonstrated. The symposium's opening session lasted for more than nine hours. During that time, many speakers stressed the uncertainties that remain about global warming and its effects—on the thermohaline circulation, on the distribution of vegetation, on the survival of cold-loving species, on the frequency of forest fires. But this sort of questioning, which is so basic to scientific discourse, never extended to the relationship between carbon dioxide and rising temperatures. The study's executive summary stated, unequivocally, that human beings had become the "dominant factor" influencing the climate. During an afternoon coffee break, I caught up with Corell.

"Let's say that there's three hundred people in this room," he told me. "I don't think you'll find five who would say that global warming is just a natural process." (While I was at the conference, I spoke to more than twenty scientists, and I couldn't find one who described it that way.)

The third part of the Arctic-climate study, which was still unfinished at the time of the symposium, was the so-called policy document. This was supposed to outline practical steps to be taken in response to the scientific findings, including—presumably—reducing greenhouse gas emissions. The policy document remained unfinished because American negotiators had rejected much of the language proposed by the seven other Arctic nations. (A few weeks later, the United States agreed to a vaguely worded statement calling for "effective"—but not obligatory—actions to combat the problem.) This recalcitrance left those Americans who had traveled to Reykjavík in an awkward position. A few tried—halfheartedly—to defend the Bush administration's stand to me; most, including many government employees, were critical of it. At one point, Corell observed that the loss of sea ice since the late 1970s was equal to "the size of Texas and Arizona combined. That analogy was made for obvious reasons."

That evening, at the hotel bar, I talked to an Inuit hunter named John Keogak, who lives on Banks Island, in Canada's Northwest Territories, some five hundred miles north of the Arctic Circle. He told me that he and his fellow hunters had started to notice that the climate was changing in the mid-eighties. Then, a few years ago, for the first time, people began to see robins, a bird for which the Inuit in his region have no word.

"We just thought, Oh, gee, it's warming up a little bit," he recalled. "It was good at the start—warmer winters, you know—but now everything is going so fast. The things that we saw coming in the early nineties, they've just multiplied.

"Of the people involved in global warming, I think we're on top of the list of who would be most affected," Keogak went on. "Our way of life, our traditions, maybe our families. Our children may not have a future. I mean, all young people, put it that way. It's just not happening in the Arctic. It's going to happen all over the world. The whole world is going too fast."

The symposium in Reykjavík lasted for four days. One morning, when the presentations on the agenda included "Char as a Model for Assessing Climate Change Impacts on Arctic Fishery Resources," I decided to rent a car and take a drive. In recent years, Reykjavík has been expanding almost on a daily basis, and the old port city is now surrounded by rings of identical, European-looking suburbs. Ten minutes from the car-rental place, these began to give out, and I found myself in a desolate landscape in which there were no trees or bushes or really even soil. The ground—fields of lava from some defunct, or perhaps just dormant, volcanoes—resembled macadam that had recently been bulldozed. I stopped to get a cup of coffee in the town of Hveragerdi, where roses are raised in greenhouses heated with steam that pours directly out of the earth. Farther on, I crossed into farm country; the landscape was still treeless, but now there was grass, and sheep eating it. Finally, I reached the sign for Sólheimajökull, the glacier whose retreat Oddur Sigurdsson had described to me. I turned off onto a dirt road. It ran alongside a brown river, between two crazily shaped ridges. After a few miles, the road ended, and the only option was to continue on foot.

By the time I got to the lookout over Sólheimajökull, it was raining. In the gloomy light, the glacier appeared less sublime than merely forlorn. Much of it was gray— covered in a film of dark grit. In its retreat, it had left behind ridged piles of silt. These were jet-black and barren—not even the tough local grasses had had a chance to take root on them. I looked around for the enormous boulder I had seen in the photos in Sigurdsson's office. It was such a long way from the edge of the glacier that for a moment I wondered if perhaps it had been carried along by the current. A raw wind came up, and I started to head down. Then I thought about what Sigurdsson had told me. If I returned in another decade, the glacier would probably no longer even be visible from the ridge where I was standing. So I climbed back up to take a second look.

Chapter 4

THE BUTTERFLY AND THE TOAD

P *OLYGONIA C-ALBUM*, generally known as the Comma butterfly, spends most of its life pretending to be something that it is not. In its larval, or caterpillar, stage, it has a chalky stripe down its back, which makes it look, uncannily, like a bird dropping. As an adult, with wings folded, it is practically indistinguishable from a dead leaf. The Comma gets its name from a tiny white mark on its underside shaped like the letter "C." Even this is thought to be part of its camouflage—an ersatz tear of the sort leaves get when they are particularly old and tatty.

The Comma is a European butterfly—its American cousins are the Eastern Comma and the Question Mark—and it can be found in France, where it is known as *le Robert-le-Diable*; Germany, where it is called *der C-Falter*, and the Netherlands, where it is *Gehakkelde Aurelia*. The Comma reaches the northern edge of distribution in England. This is unremarkable—many European butterflies come to the end of their range in Britain—but from a scientific standpoint fortunate.

The English have been watching and collecting butterflies for centuries—some of the specimens in the British Natural History Museum date back to the 1700s—and in the Victorian era, passion for the hobby was such that every city, and many a small town, supported its own entomological society. In the 1970s, Britain's Biological Records Centre decided to marshal this enthusiasm for a project called the Lepidoptera Distribution Maps Scheme, whose aim was to chart precisely where each of the country's fifty-nine native species could—and could not—be found. More than two thousand amateur lepidopterists participated, and in 1984, the results were collated into a hundred-and-fifty-eight-page atlas. Every species got its own map with different colored dots showing the number of times it had been sighted in any given ten square kilometers. In the map for *Polygonia c-album*, the Comma's range was shown to extend from the south coast of England northward to Liverpool in the west and Norfolk in the east. Almost immediately, this map became out of date; in the years that followed, hobbyists kept finding the Comma in new areas. By the late 1990s, the butterfly was frequently being sighted in the north of England, near Durham. By now it is well established in southern Scotland, and has been sighted as far north as the Scottish Highlands. The rate of the Comma's expansion— some fifty miles per decade—was described by the authors of the most recent butterfly atlas as "remarkable."

Chris Thomas is a biologist at the University of York who studies lepidoptera. He is tall and rangy, with an Ethan Hawke–style goatee and an amiably harried manner.

ique—conditions. These include the Chalkhill Blue (*Polyommatus coridon*), a large, turquoise butterfly that feeds exclusively on horseshoe vetch, and the Purple Emperor (*Apatura iris*), which flies in the treetops of well-wooded areas in southern England. Then there are the "generalists," who are less picky. Among Britain's generalists, there are, in addition to the Comma, ten species that are widespread in the southern part of the country and reach the edge of their range somewhere in the nation's midsection. "Every single one has moved northward since 1982," Thomas told me. A few years ago, together with lepidopterists from, among other places, the United States, Sweden, France, and Estonia, Thomas conducted a survey of all the studies that had been done on generalists that reach the upper limits of their ranges in Europe. The survey looked at thirty-five species in all. Of these, the scientists found, twenty-two had shifted their range northward in recent decades, while only one had shifted south.

After a while, the sun emerged, and we went back outside. Thomas's wolfhound, Rex, a dog the size of a small horse, trailed behind us, panting heavily. Within about five minutes, Thomas had identified a Meadow Brown (*Maniola jurtina*), a Small Tortoiseshell (*Aglais urticae*), and a Green-veined White (*Pieris napi*), all species that have been flitting around Yorkshire since butterfly record-keeping began. Thomas also spotted a Gatekeeper (*Pyronia tithonus*) and a Small Skipper (*Thymelicus sylvestris*), which until recently had been confined to a region well south of where we were standing. "So far, two out of the five species of butterflies that we've seen are northward invaders," he told me. "Sometime within the last thirty years they have spread into this area." A few minutes later, he pointed out another invader sunning itself in the grass—a *Polygonia c-album*. With its wings closed, the Comma was a dull, dead-leaf brown, but with them open, it was a brilliant orange.

That life on earth changes with the climate has been assumed to be the case for a long time, indeed for very nearly as long as the climate has been known to be capable of changing. Louis Agassiz published *Études sur les glaciers*, the work in which he laid out his theory of the ice ages, in 1840. By 1859, Charles Darwin had incorporated Agassiz's theory into his own theory of evolution. Toward the end of *On the Origin of Species*, in a chapter titled "Geographical Distribution," Darwin describes the vast migrations that he supposes the advance and retreat of the glaciers must have necessitated:

> As the cold came on, and as each more southern zone became fitted for arctic beings and ill-fitted for their former more temperate inhabitants, the latter would be supplanted and arctic productions would take their places. The inhabitants of the more temperate regions would at the same time travel southward . . . As the warmth returned, the arctic forms would retreat north-ward, closely followed up in their retreat by the pro-ductions of the more temperate regions. And as the snow melted from the bases of the mountains, the arctic forms would seize on the cleared and thawed ground,

always ascending higher and higher as the warmth increased, whilst their brethren were pursuing their northern journey.

For Darwin and his contemporaries such a narrative was necessarily speculative. Much as the existence of ice ages had had to be inferred from the signs they left behind— erratics, moraines, and striated bedrock—so, too, the succession and redistribution of species on Earth could only be reconstructed from fragmentary traces: scattered bones, fossilized insect casings, ancient pollen deposits. Even as paleontologists and paleobotanists found more and more evidence of how species had responded to climate change in the past, it was taken for granted that the process was not something that could be observed in real time, an assumption that has now been proven false.

Almost anywhere you go in the world today, except perhaps for the urban areas where most of us live, it is possible to observe biological changes comparable to the northern expansion of the Comma. A recent study of common frogs living near Ithaca, New York, for example, found that four out of six species were calling—which is to say, mating—at least ten days earlier than they used to, while at the Arnold Arboretum, in Boston, the date of peak blooming for spring-flowering shrubs has advanced, on average, by eight days. In Costa Rica, birds like the keel-billed toucan (*Ramphastos sulfuratus*), once confined to the lowlands, have started to nest on mountain slopes; in the Alps, plants like purple saxifrage (*Saxifraga oppositifolia*) and Austrian draba (*Draba fladnizensis*) have been creeping up toward the summits; and in the Sierra Nevada of California, the average Edith's Checkerspot butterfly (*Euphydryas editha*) can now be found at an elevation three hundred feet higher than it was a hundred years ago. Any one of these changes could, potentially, be a response to purely local conditions—shifts, say, in regional weather patterns or in patterns of land use. The only explanation that anyone has proposed that makes sense of them all, though, is global warming.

The Bradshaw-Holzapfel Lab occupies a corner on the third floor of Pacific Hall, a peculiarly unlovely building on the campus of the University of Oregon in Eugene. At one end of the lab is a large room stacked with glassware and at the other, a pair of offices. In between are several workrooms that look, from the outside, like walk-in refrigerators. Taped to the door of one of them is a handwritten sign: "Warning—if you enter this room mosquitoes will suck your blood out through your eyes!"

William Bradshaw and Christina Holzapfel, who run the lab and share one of the offices, are evolutionary biologists. They were introduced as graduate students at the University of Michigan, and have been married for thirty-five years. Bradshaw is a tall man with thinning gray hair and a gravelly voice. His desk is covered in a mess of papers, books, and journals, and when visitors come to the lab, he likes to show them his collection of curiosities, which includes a desiccated octopus. Holzapfel is short, with blond hair and bright blue eyes. Her desk is perfectly neat.

Bradshaw and Holzapfel have shared an interest in mosquitoes for as long as they have been interested in each other. In the early years of their lab, which they set up in 1971, they raised several species, some of which, in order to reproduce, required what is delicately referred to as a "blood meal." This, in turn, demanded a live animal able to provide such a meal. For a time, this requirement was met by rats sedated with phenobarbital, but, as rules about experimenting with animals grew more stringent, Bradshaw and Holzapfel found themselves forced to decide whether it was more humane to keep sedating the same rat over and over again, or to use a new rat and let the old one wake up to find itself covered with bites. Eventually, they grew weary of such questions and decided to stick to a single species, *Wyeomyia smithii*, which needs no blood in order to reproduce. At any given moment the Bradshaw-Holzapfel Lab houses upwards of a hundred thousand *Wyeomyia smithii* in various stages of development.

Wyeomyia smithii is a small and rather ineffectual bug. ("Wimpy" is how Bradshaw characterizes it.) Its eggs are practically indistinguishable from specks of dust; its larvae appear as minuscule white worms. As an adult, it is about a quarter of an inch long and in flight looks like a tiny black blur. It is only when you examine a *Wyeomyia smithii* very closely, under a magnifying glass, that you can see that its abdomen is actually silver, and that its two hind legs are bent gracefully above its head, like a trapeze artist's.

Wyeomyia smithii completes virtually its entire life cycle—from egg to larva to pupa to adult—inside a single plant, *Sarracenia purpurea* or, as it is more commonly known, the purple pitcher plant. The purple pitcher plant, which grows in swamps and peat bogs from Florida to northern Canada, has frilly, cornucopia-shaped leaves that sprout directly out of the ground and then fill with water. In the spring, female *Wyeomyia smithii* lay their eggs one at a time, carefully depositing each in a different pitcher plant. When flies and ants and occasionally small frogs drown in the leaves of the pitcher plant—*Sarracenia purpurea* is carnivorous—their remains also provide nutrients for developing mosquito larvae. (*Sarracenia purpurea* does not digest its own food; it leaves this task to bacteria, which don't attack the mosquitoes.) When the young mature into adults, they repeat the whole process, and if conditions are favorable, the cycle can be completed four or five times in a single summer. Come fall, the adult mosquitoes die off, but the larvae live on through the winter in a state known as diapause—the insect version of hibernation.

The exact timing of diapause is critical to the survival of *Wyeomyia smithii* and also to Bradshaw and Holzapfel's research. In contrast to most insects, which rely on a variety of signals, including temperature and food availability, to regulate the onset of dormancy, *Wyeomyia smithii* depends exclusively on light cues. When the larvae perceive that day length has dropped below a certain threshold, they stop growing and molting; when they perceive that it has lengthened sufficiently, they take up again where they left off.

This light threshold, which is known as the critical photoperiod, varies from bog to bog. At the southern end of the mosquitoes' range, near the Gulf of Mexico, conditions remain favorable for breeding well into fall. A typical *Wyeomyia smithii* from Florida or Alabama will, consequently, not go dormant until day length has shrunk to about twelve and a half hours, which at that latitude corresponds to early November. At the far northern edge of the range, meanwhile, winter arrives much earlier, and an average mosquito from Manitoba will go into dormancy in late July, as soon as day length drops below sixteen and a half hours. Interpreting light cues is a genetically controlled and highly heritable trait: *Wyeomyia smithii* are programmed to respond to day length the same way their parents did, even if they find themselves living under very different conditions. (One of the walk-in-freezer-like rooms in the Bradshaw-Holzapfel Lab contains locker-size storage units, each equipped with a timer and a fluorescent bulb, where mosquito larvae can be raised under any imaginable schedule of lightness and dark.) In the mid-1970s, Bradshaw and Holzapfel demonstrated that *Wyeomyia smithii* living at different elevations also obey different light cues—high-altitude mosquitoes behave as if they were born farther north—a discovery that today might seem relatively unremarkable but at the time was sufficiently noteworthy to make the cover of *Nature*.

About five years ago, Bradshaw and Holzapfel began to

wonder about how *Wyeomyia smithii* might be affected by global warming. They knew that the species had expanded northward after the end of the last glaciation, and that at some point in the intervening millennia, the critical photoperiods of northern and southern populations had diverged. If climatic conditions were changing once again, then perhaps this would show up in the timing of diapause. The first thing the couple did was go back to look at their old data, to see if it contained any information that they hadn't previously noticed.

"There it was," Holzapfel told me. "Just hitting you right in the eye."

When an animal changes its routine by, say, laying its eggs earlier or going into hibernation later, there are a number of possible explanations. One is that the change reflects an innate flexibility; as conditions vary, the animal is able to adjust its behavior in response. Biologists call such flexibility "phenotypic plasticity," and it is key to the survival of most species. Another possibility is that the shift represents something deeper and more permanent—an actual rearrangement of the organism's genetic code.

In the years since they established their lab, Bradshaw and Holzapfel have collected mosquito larvae from all over the eastern United States and much of Canada. The couple used to do the collecting themselves, driving across the country in a van equipped with a makeshift bed for their daughter and a miniature lab for sorting, labeling, and storing the thousands of specimens they would gather. Nowadays, they more often send out their graduate students, who, instead of driving, are likely to fly. (Getting through airport security with a backpack full of mosquito larvae is a process that, the students have learned, can take half a day.)

Every subpopulation exhibits a range of light responses; Bradshaw and Holzapfel define critical photoperiod as the point at which 50 percent of the mosquitoes in a sample have switched from active development to diapause. Each time they collect a new batch of insects, they put the larvae in petri dishes and place the dishes in the controlled-environment light boxes, which they call Mosquito Hiltons. Then they test the larvae for their critical photoperiod, and record the results.

When Bradshaw and Holzapfel went back to their files, they looked for populations that they had tested at least twice. One of these was from a wetland called Horse Cove, in Macon County, North Carolina. In 1972, when the couple had collected mosquitoes for the first time from Horse Cove, their files showed, the larvae's critical photoperiod was fourteen hours and twenty-one minutes. They collected a second batch of mosquitoes from the same spot in 1996. By that point, the insects' critical photoperiod had dropped to thirteen hours and fifty-three minutes. All told, Bradshaw and Holzapfel found that in their files they had comparative data on ten different subpopulations—two in Florida, three in North Carolina, two in New Jersey, and one each in Alabama, Maine, and Ontario. In every single case, the critical photoperiod had declined over time. Also, their data showed that the farther north you went, the stronger the effect; a regression analysis revealed that the critical photo-period of mosquitoes living at fifty degrees north latitude had declined by more than thirty-five minutes, corresponding to a delay in diapause of nearly nine days.

In a different mosquito, this shift could be an instance of the kind of plasticity that allows organisms to cope with varying conditions. But in *Wyeomyia smithii*, there is no flexibility when it comes to timing the onset of diapause. Warm or cold, all the insect can do is read light. Bradshaw and Holzapfel knew therefore that the change they were seeing must be genetic. As the climate had warmed, those mosquitoes that had remained active until later in the fall had enjoyed a selective advantage, presumably because they had been able to store a few more days' worth of resources for the winter, and they had passed this advantage on to their offspring, and so on. In December 2001, Bradshaw and Holzapfel published their findings in the

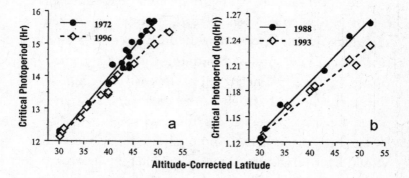

The critical photoperiod for Wyeomyia smithii *has declined markedly over time. Changes are most dramatic at higher latitudes. Credit: After W. Bradshaw and C. Holzapfel,* PNAS, *vol. 98 (2001).*

Proceedings of the National Academy of Sciences. By doing so, they became the first researchers to demonstrate that global warming had begun to drive evolution.

The Monteverde Cloud Forest sits astride the Cordillera de Tilarán, or Tilarán Mountains, in north-central Costa Rica. The rugged terrain in combination with the trade winds that blow off the Caribbean Sea make the region unusually diverse; in an area of less than two hundred and fifty square miles, there are seven "life zones," each with its own distinctive type of vegetation. The cloud forest is surrounded on all sides by land, yet, ecologically speaking, it is an island and, as is often the case with islands, it displays a high degree of endemism, or biological specificity. Fully 10 percent of Monteverdean flora, for example, are believed to be unique to the area.

The most famous of Monteverde's endemic species is—or at least was—a small toad. Known colloquially as the golden toad, it was officially discovered by a biologist from the University of Southern California named Jay Savage. Savage had heard tell of the toad from a group of Quakers who had settled at the edge of the forest; still, when he came across it for the first time, on May 14, 1964, at the top of a high mountain ridge, his reaction, he would later recall, was "one of disbelief." Most toads are dull brown, grayish green, or olive; this one was a flaming shade of tangerine. Savage named the new species *Bufo periglenes*, from the Greek word meaning bright, and titled his paper on the discovery "An Extraordinary New Toad (Bufo) from Costa Rica."

Since the golden toad spent its life underground, emerging only in order to reproduce, most of what was subsequently learned about it had to do with sex. The toad was, it was determined, an "explosive breeder"; instead of staking out and defending territory, males simply rushed the first available female and fought for the chance to mount her. ("Amplexus" is the term of art for an amphibian embrace.) Males outnumbered females, in some years by as much as ten to one, a situation that often led bachelors to attack amplectant pairs and form what Savage once described as "writhing masses of toad balls." The eggs of the golden toad, black and tan spheres, were deposited in small pools—puddles, really—often no more than one inch deep. Tadpoles emerged in a matter of days, but required another four or five weeks for metamorphosis. During this period, they were highly dependent on the weather; too much rain and they would be washed down the steep hillsides, too little and their puddles would dry up. Golden toads were never found more than a few miles from the site where Savage originally spotted them, always at the top of a mountain ridge, and always at an altitude of between forty-nine hundred and fifty-six hundred feet.

In the spring of 1987, an American biologist who had come to the cloud forest specifically to study the toads counted fifteen hundred of them in temporary breeding pools. That spring was unusually warm and dry, and most of the pools evaporated before the tadpoles in them had

had time to mature. The following year, only one male was seen at what previously had been the major breeding site. Seven males and two females were seen at a second site a few miles away. The year after that, a search of all spots where the toad had earlier been sighted yielded a solitary male. No golden toad has been seen since, and it is widely assumed that after living its colorful, if secretive, existence for hundreds of thousands of years, *Bufo periglenes* is now extinct.

In April 1999, J. Alan Pounds, who heads the Golden Toad Laboratory for Conservation in the Monteverde Preserve, published a paper in *Nature* on the toad's demise. In it, he linked the toad's extinction, as well as the decline of several other amphibian species, to a shift in precipitation patterns in the cloud forest. In recent years, there has been a significant increase in the number of days with no measurable precipitation, a change that, in turn, is consonant with an increase in the elevation of the cloud cover. In a separate article in the same issue of *Nature*, a group of scientists from Stanford University reported on their efforts to model the future of cloud forests. They predicted that as global CO_2 levels continued to rise, the height of the cloud cover in the Monteverde Preserve and other tropical cloud forests would continue to climb. This, they speculated, would force a growing number of high-altitude species "out of existence."

Climate change – even violent climate change – is itself, of course, part of the natural order. For the earth's flora, the last two million years have been particularly turbulent; in addition to the glacial cycles, there have also been dozens of abrupt climate shifts, like the Younger Dryas.

Thompson Webb III is a paleoecologist who teaches at Brown University. He studies pollen grains and fern spores, in an effort to reconstruct the plant life of previous eras. In the mid-seventies, Webb began to assemble a database of pollen records from lakes all across North America. (When a grain of pollen falls on the ground, it usually oxidizes and disappears; if it is blown onto a body of water, however, it can sink to the bottom and be preserved in the sediment for millennia.) The project took nearly twenty years to complete, and, when it was finally done, it showed how, as the climate of the continent had changed, life had rearranged itself.

A few months after I visited Bill Bradshaw and Chris Holzapfel in Eugene, I went to talk to Webb in Providence. He has an office in the university's geochemistry building, and also a lab, where, on this particular day, one of his research assistants was examining charcoal particles from an ancient forest fire. Webb took some slides from a cabinet and slipped one under the lens of a microscope. Most pollen grains are between twenty and seventy microns in diameter; to be identified, they must be magnified four hundred times. Peering through the eyepiece, I saw a tiny sphere, pocked like a golf ball. Webb told me that what I was looking at was a grain of birch pollen. He replaced the slide, and a second tiny golf ball swam into focus. It was beech pollen, Webb explained, and could be distinguished by a set of three minute grooves. "You see, they're really very different," he said of the two grains.

After a while, we went down the hall to Webb's office. On his computer he called up a program named Pollen Viewer 3.2, and a map of North America circa 19000 B.C. appeared on the screen. Around that time, the ice sheets of the last glaciation reached their maximum extent; the map showed the Laurentide ice sheet covering all of Canada as well as most of New England and the upper Midwest. Because so much water was tied up in the ice, sea levels were some three hundred feet lower than they are now. On the map, Florida appeared as a stubby protuberance, nearly twice as wide as it is today. Webb clicked on "Play." Time began to move forward in thousand-year increments. The ice sheet shrank. A huge lake, known as Lake Agassiz, formed in central Canada and, a few thousand years later, drained. The Great Lakes emerged, and then widened. Around eight thousand years ago, open water finally appeared in Hudson Bay. The bay began to contract as the land around it rebounded from the weight of the ice sheet.

Webb clicked on a pull-down menu that listed the Latin names of dozens of trees and shrubs. He chose *Pinus* (pine) and again hit "Play." Dark-green splotches began to move around the continent. Twenty-one thousand years ago, the program showed, pine forests covered the entire Eastern Seaboard south of the ice sheet. Ten thousand years later, pines were concentrated around the Great Lakes, and today pine predominates in the southeastern United States

THE BUTTERFLY AND THE TOAD

and in western Canada. Webb clicked on *Quercus* (oak), and a similar process began, only *Quercus* moved in a very different pattern from *Pinus*. More clicks for *Fagus* (beech), *Betula* (birch), and *Picea* (spruce). As the earth warmed and the continent emerged from the ice age, each of the tree species migrated, but no two moved in exactly the same way.

"The trick you've got to remember is that climate is multivariate," Webb explained. "The plant species are having to respond both to temperature changes and to moisture changes and to changes in seasonality. It makes a big difference if you have a drier winter versus a drier summer, because some species are more attuned to spring and others to fall. Any current community has a certain mixture. If you start changing the climate, you're changing the temperature, but you're also changing moisture or the timing of the moisture or the amount of snow and, bingo, species are not going to move together. They can't."

Webb pointed out that the warming predicted for the next century is on the same scale as the temperature difference between the last glaciation and today. "You know that's going to give us a very different landscape," he said. I asked what he thought this landscape would look like. He said he didn't know—his central finding, from more than thirty years of research, is that, as the climate changes, species often move in surprising ways. In the short term, which is to say in the remainder of his own life, Webb said that he expected mostly to see disruption.

"We have this strange sense of the evolutionary hierarchy, that the microorganisms, because they came first, are the most primitive," he told me. "And yet you could argue that this will just give a lot of advantage to the microorganisms of the world, because of their ability to evolve more quickly. To the extent the climate is putting organisms as well as ecosystems under stress, it's opening the opportunities for invasive species on the one hand and disease on the other. I guess I start thinking: Think death."

Any species that is around today, including our own, has already survived catastrophic climate change. The fact that a species has survived such a change, or even many such changes, is no guarantee, however, that it will survive the next one. Consider, for example, the outsized megafauna— seven-hundred-and-fifty-pound saber-toothed cats, elephantine sloths, and fifteen-foot-tall mastodons—that once dominated the North American landscape. These megafauna lived through several glacial cycles, but then something changed, and they nearly all died out at the same time, at the beginning of the Holocene.

Over the past two million years, even as the temperature of the earth has swung wildly, it has always remained within certain limits: The planet has often been colder than today, but rarely warmer, and then only slightly. If the earth continues to warm at the current rate, then by the end of this century temperatures will push beyond the "envelope" of natural climate variability.

Meanwhile, thanks to us, the world today is a very different—and in many ways diminished—place. International trade has introduced exotic pests and competitors; ozone depletion has increased exposure to ultraviolet radiation; and many species have already been very nearly wiped out, or wiped out altogether, by overhunting and overharvesting. Perhaps most significantly, human activity, in the form of farms and cities and subdivisions and mines and logging operations and parking lots, has steadily reduced the amount of available habitat. G. Russell Coope is a visiting professor in the geography department at the University of London and one of the world's leading authorities on ancient beetles. He has shown that, under the pressure of climate change, insects have migrated tremendous distances; for example, *Tachinus caelatus*, a small, dullish-brown beetle common in England during the cold periods of the Pleistocene, today can be found only some five thousand miles away, in the mountains west of Ulan Bator, in Mongolia. But Coope questions whether such long-distance migrations are practical in a fragmented landscape like today's. Many organisms now live in the functional equivalent of "oceanic islands or remote mountain tops," he has written. "Certainly, our knowledge of their past response may be of little value in predicting any future reactions to climate change, since we have imposed totally new restrictions on their mobility; we have inconveniently moved the goal posts and set up a ball game with totally new rules."

A few years ago, nineteen biologists from around the world set out to give, in their words, a "first pass" estimate of the extinction risk posed by global warming. They assembled data on eleven hundred species of plants and animals from sample regions covering roughly a fifth of the earth's surface. Then they established the species' current ranges, based on climate variables such as temperature and rainfall. Finally, they calculated how much of the species' "climate envelope" would be left under different warming scenarios. The results of this effort were published in *Nature* in 2004. Using a midrange projection of temperature rise, the biologists concluded that, if the species in the sample regions could be assumed to be highly mobile, then fully 15 percent of them would be "committed to extinction" by the middle of this century, and, if they proved to be basically stationary, an extraordinary 37 per cent of them would be.

The Mountain Ringlet (*Erebia epiphron*) is a dun-colored butterfly with orange and black spots that curl along the edges of its rounded wings. Mountain Ringlets feed on a coarse, tufted grass known as matgrass, overwinter as larvae, and as adults have an extremely brief lifespan— perhaps as short as one or two days. A montane, or mountain species, it is found only at elevations above a thousand feet in the Scottish Highlands, and farther south, in Britain's Lake District, only above fifteen hundred feet.

Together with a colleague from the University of York, Chris Thomas has for the last few years been monitoring the Mountain Ringlet, along with three other species of butterfly—the Scotch Argus (*Erebia aethiops*), the Large Heath (*Coenonympha tullia*) and the Northern Brown

Argus (*Aricia artaxerxes*)—whose ranges are similarly confined to a few locations in northern England and Scotland. In the summer of 2004, researchers for the project visited nearly six hundred sites where these "specialist" species had been spotted in the past, and the following summer they repeated the process. Documenting a species' contraction is more difficult than documenting its expansion—is it really gone, or did someone just miss it?—but preliminary evidence suggests that the butterflies are already disappearing from lower elevation, and therefore warmer, sites. When I went to visit Thomas, he was getting ready to take his family to Scotland on vacation, and was planning to recheck some of the sites. "It's a bit of a busman's holiday," he confessed.

As we were wandering around his yard in search of Commas, I asked Thomas, who was the lead author of the extinction study, how he felt about the changes he was seeing. He told me that he found the opportunities for study presented by climate change to be exciting.

"Ecology for a very long time has been trying to explain why species have the distribution that they do, why a species can survive here and not over there, why some species have small distributions and others have broad ones," he said. "And the problem that we have always had is that distributions have been rather static. We couldn't actually see the process of range boundaries changing taking place, or see what was driving those changes. Once everything starts moving, we can begin to understand: is it a climatic determinant, or is it mainly other things, like interactions with other species? And, of course, if you think of the history of the last million years, we now have the opportunity to try and understand how things might have responded in the past. It's extremely interesting, the prospect of everything changing its distribution, and new mixtures of species from around the world starting to form and produce new biological communities—extremely interesting from a purely academic point of view.

"On the other hand, given our conclusions about possible extinctions, it is, to me personally, a serious concern," he went on. "If we are in the situation where a quarter of the terrestrial species might be at risk of extinction from climate change—people often use the phrase 'being like canaries'—if we've changed our biological system to such an extent, then we do have to get worried about whether the services that are provided by natural ecosystems are going to continue. Ultimately, all of the crops we grow are biological species; all the diseases we have are biological species; all the disease vectors are biological species. If there is this overwhelming evidence that species are changing their distributions, we're going to have to expect exactly the same for crops and pests and diseases. Part of it simply is we've got one planet, and we are heading it in a direction that, quite fundamentally, we don't know what the consequences are going to be."

Part II

MAN

Chapter 5

THE CURSE OF AKKAD

THE WORLD'S FIRST empire was established forty-three hundred years ago, between the Tigris and Euphrates Rivers. The details of its founding, by Sargon of Akkad, have come down to us in a form somewhere between history and myth. Sargon—Sharru-kin, in the Akkadian language—means "true king"; almost certainly, though, he was a usurper. As a baby, Sargon was said to have been discovered, Moses-like, floating in a basket. Later, he became cupbearer to the ruler of Kish, one of ancient Babylonia's most powerful cities. Sargon dreamed that his master, Ur-Zababa, was about to be drowned by the goddess Inanna in a river of blood. Hearing about the dream, Ur-Zababa decided to have Sargon eliminated. How this plan failed is unknown; no text relating the end of the story has ever been found.

Until Sargon's reign, Babylonian cities like Kish, and also Ur and Uruk and Umma, functioned as independent city-states. Sometimes they formed brief alliances—cuneiform tablets attest to strategic marriages celebrated and diplomatic gifts exchanged—but mostly they seem to have been at war with one another. Sargon first subdued Babylonia's fractious cities, then went on to conquer, or at least sack, lands like Elam, in present-day Iran. He presided over his empire from the city of Akkad, the ruins of which are believed to lie south of Baghdad. It was written that "daily five thousand four hundred men ate at his presence," meaning, presumably, that he maintained a huge standing army. Eventually, Akkadian hegemony extended as far as the Khabur plains, in northeastern Syria, an area prized for its grain production. Sargon came to be known as "king of the world"; later, one of his descendants enlarged this title to "king of the four corners of the universe."

Akkadian rule was highly centralized, and in this way anticipated the administrative logic of empires to come. The Akkadians levied taxes, then used the proceeds to support a vast network of local bureaucrats. They introduced standardized weights and measures—the *gur* equalled roughly three hundred liters—and imposed a uniform dating system, under which each year was assigned the name of a major event that had recently occurred: for instance, "the year that Sargon destroyed the city of Mari." Such was the level of systematization that even the shape and the layout of accounting tablets were imperially prescribed. Akkad's wealth was reflected in, among other things, its artwork, the refinement and naturalism of which were unprecedented.

Sargon ruled, supposedly, for fifty-six years. He was succeeded by his two sons, who reigned for a total of twenty-four years, and then by a grandson, Naram-sin, who declared himself a god. Naram-sin was, in turn, succeeded by his son. Then, suddenly, Akkad collapsed. During one three-year period, four men each, briefly, claimed the throne. "Who was king? Who was not king?" the register known as the Sumerian King List asks, in what may be the first recorded instance of political irony.

The lamentation "The Curse of Akkad" was written within a century of the empire's fall. It attributes Akkad's demise to an outrage against the gods. Angered by a pair of inauspicious oracles, Naram-sin plunders the temple of Enlil, the god of wind and storms, who, in retaliation, decides to destroy both him and his people:

> For the first time since cities were built and founded,
> The great agricultural tracts produced no grain,
> The inundated tracts produced no fish,
> The irrigated orchards produced neither syrup nor
> wine,
> The gathered clouds did not rain, the *masgurum* did not
> grow.
> At that time, one shekel's worth of oil was only one-half
> quart,
> One shekel's worth of grain was only one-half quart . . .
> These sold at such prices in the markets of all the cities!
> He who slept on the roof, died on the roof,
> He who slept in the house, had no burial,
> People were flailing at themselves from hunger.

For many years, the events described in "The Curse of Akkad" were thought, like the details of Sargon's birth, to be purely fictional.

In 1978, after scanning a set of maps at Yale's Sterling Memorial Library, a university archaeologist named Harvey Weiss spotted a promising-looking mound at the confluence of two dry riverbeds in the Khabur plains, near the Iraqi border. He approached the Syrian government for permission to excavate the mound, and, somewhat to his surprise, it was almost immediately granted. Soon, he had uncovered a lost city, which in ancient times was known as Shekhna and today is called Tell Leilan.

Over the next ten years, Weiss, working with a team of students and local laborers, proceeded to uncover an acropolis, a crowded residential neighborhood reached by a paved road, and a large block of grain-storage rooms. He found that the residents of Tell Leilan had raised barley and several varieties of wheat, that they had used carts to transport their crops, and that in their writing they had imitated the style of their more sophisticated neighbors to the south. Like most cities in the region at the time, Tell Leilan had a rigidly organized, state-run economy: people received rations—so many liters of barley and so many of oil—based on how old they were and what kind of work they performed. From the time of the Akkadian empire, thousands of similar potsherds were discovered, indicating that residents had received their rations in mass-produced, one-liter vessels. After examining these and other artifacts,

Weiss constructed a time line of the city's history, from its origins as a small farming village (around 5000 B.C.), to its growth into an independent city of some thirty thousand people (2600 B.C.), and on to its reorganization under imperial rule (2300 B.C.).

Wherever Weiss and his team dug, they also encountered a layer of dirt that contained no signs of human habitation. This layer, which was more than three feet deep, corresponded to the years 2200 to 1900 B.C., and it indicated that, around the time of Akkad's fall, Tell Leilan had been completely abandoned. In 1991, Weiss sent soil samples from Tell Leilan to a lab for analysis. The results showed that, around the year 2200 B.C., even the city's earthworms had died out. Eventually, Weiss came to believe that the lifeless soil of Tell Leilan and the end of the Akkadian empire were products of the same phenomenon—a drought so prolonged and so severe that it represented, in his words, an example of "climate change."

Weiss first published his theory, in the journal *Science*, in August 1993. Since then, the list of cultures whose demise has been linked to climate change has continued to grow. They include the Classic Mayan civilization, which collapsed at the height of its development, around A.D. 800; the Tiwanaku civilization, which thrived near Lake Titicaca, in the Andes, for more than a millennium, then disintegrated around A.D. 1100; and the Old Kingdom of Egypt, which collapsed around the same time as the Akkadian empire. (In an account eerily reminiscent of "The Curse of Akkad," the Egyptian sage Ipuwer described the anguish of the period: "Lo, the desert claims the land. Towns are ravaged . . . Food is lacking . . . Ladies suffer like maidservants. Lo, those who were entombed are cast on high grounds.") In each of these cases, what began as a provocative hypothesis has, as new information has emerged, come to seem more and more compelling. For example, the notion that Mayan civilization had been undermined by climate change was first proposed in the late 1980s, at which point there was little climatological evidence to support it. Then, in the mid-1990s, American scientists studying sediment cores from Lake Chichancanab, in north-central Yucatán, reported that precipitation patterns in the region had indeed shifted during the ninth and tenth centuries, and that this shift had led to periods of prolonged drought. More recently, a group of researchers examining ocean-sediment cores collected off the coast of Venezuela produced an even more detailed record of rainfall in the area. They found that the region experienced a series of severe, "multiyear drought events" beginning around A.D. 750. The collapse of the Classic Mayan civilization, which has been described as "a demographic disaster as profound as any other in human history," is thought to have cost several million lives.

The climate shifts that affected past cultures predate industrialization by hundreds—or, in some cases, thousands—of years. They reflect the climate system's innate variability and could not have been foreseen by the societies that experienced them. Caught by surprise, the Akkadians made sense of their suffering as divine retribution. The climate shifts predicted for the coming century, by contrast, are attributable to forces whose causes we know and whose magnitude we will determine.

The Goddard Institute for Space Studies, or GISS, is situated just south of Columbia University's main campus, at the corner of Broadway and West 112th Street. The institute is not well marked, but most New Yorkers would probably recognize the building: its ground floor is home to Tom's Restaurant, the coffee shop made famous by *Seinfeld*.

GISS, an outpost of NASA, started out, forty-five years ago, as a planetary-research center; today, its major function is making climate forecasts. GISS employs about a hundred and fifty people, many of whom spend their days working on calculations that may—or may not—end up being incorporated in the institute's climate model. Some work on algorithms that describe the behavior of the atmosphere, some on the behavior of the oceans, some on vegetation, some on clouds, and some on making sure that all these algorithms, when they are combined, produce results that seem consistent with the real world. (Once, when some refinements were made to the model, rain nearly stopped falling over the world's rainforests.) The latest version of the GISS model, called ModelE, consists of 125,000 lines of computer code.

GISS's director, James Hansen, occupies a spacious, almost comically cluttered office on the institute's seventh floor. (I must have expressed some uneasiness the first time

I visited him, because the following day I received an e-mail assuring me that the office was "a lot better organized than it used to be.") Hansen, who is sixty-three, is a spare man with a lean face and a fringe of brown hair. Although he has probably done as much to publicize the dangers of global warming as any other scientist, in person he is reticent almost to the point of shyness. When I asked him how he had come to play such a prominent role, he just shrugged. "Circumstances," he said.

Hansen first became interested in climate change in the mid-1970s. Under the direction of James Van Allen (for whom the Van Allen radiation belts are named), he had written his doctoral dissertation on the climate of Venus. In it, he had proposed that the planet, which has an average surface temperature of 876 degrees Fahrenheit, was kept warm by a smoggy haze; soon afterward, a space probe showed that Venus was actually insulated by an atmosphere that consists of 96 percent carbon dioxide. When solid data began to show what was happening to greenhouse gas levels on Earth, Hansen became, in his words, "captivated." He decided that a planet whose atmosphere could change in the course of a human lifetime was more interesting than one that was going to continue, for all intents and purposes, to broil away forever. A group of scientists at NASA had put together a computer program to try to improve weather forecasting using satellite data. Hansen and a team of half a dozen other researchers set out to modify it, in order to make longer-range forecasts about what would happen to global temperatures as greenhouse

gases continued to accumulate. The project, which resulted in the first version of the GISS climate model, took
nearly seven years to complete.

At that time, there was little empirical evidence to
support the notion that the earth was warming. Instrumental temperature records go back, in a consistent fash-
ion, only to the mid-nineteenth century. They show that
average global temperatures rose through the first half of
the twentieth century, then dipped in the 1950s and '60s.
Nevertheless, by the early 1980s Hansen had gained enough confidence in his model to begin to make a series of
increasingly audacious predictions. In 1981, he forecast that
"carbon dioxide warming should emerge from the noise of
natural climate variability" around the year 2000. During
the exceptionally hot summer of 1988, he appeared before
a Senate subcommittee and announced that he was "99
percent" sure that "global warming is affecting our planet
now." And in the summer of 1990 he offered to bet a
roomful of fellow scientists a hundred dollars that either
that year or one of the following two years would be the
warmest on record. To qualify, the year would have to set
a record not only for land temperatures but also for sea-
surface temperatures and for temperatures in the lower
atmosphere. Hansen won the bet in six months.

Like all climate models, GISS's divides the world into a
series of boxes. Thirty-three hundred and twelve boxes
cover the earth's surface, and this pattern is repeated
twenty times moving up through the atmosphere, so that

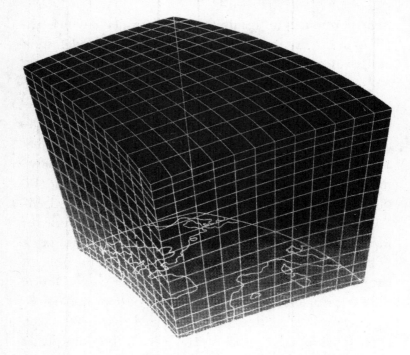

Climate models divide the world into a series of boxes. Credit: Global Warming: The Complete Briefing, *Cambridge University Press.*

the whole arrangement might be thought of as a set of enormous checkerboards stacked on top of one another. Each box represents an area of four degrees latitude by five degrees longitude. (The height of the box varies depending on altitude.) In the real world, of course, such a large area would have an incalculable number of features; in the world of the model, features such as lakes and forests and, indeed, whole mountain ranges are reduced to a limited set of properties, which are then expressed as numerical approximations. Time in this grid-world moves ahead for the most part in discrete, half-hour intervals, meaning that a new set of calculations is performed for each box for every thirty minutes that is supposed to have elapsed in actuality. Depending on what part of the globe a box represents, these calculations may involve dozens of different algorithms, so that a model run that is supposed to simulate climate conditions over the next hundred years involves more than a quadrillion separate operations. A single run of the GISS model, done on a supercomputer, usually takes about a month.

Very broadly speaking, there are two types of equations that go into a climate model. The first group expresses fundamental physical principles, like the conservation of energy and the law of gravity. The second group describes—the term of art is "parameterize"—patterns and interactions that have been observed in nature but may be only partly understood, or processes that occur on a small scale and have to be averaged out over huge spaces. Here, for example, is a tiny piece of ModelE, written in the computer language FORTRAN, which deals with the formation of clouds:

```
C**** COMPUTE THE AUTOCONVERSION RATE OF CLOUD WATER TO
PRECIPITATION
   RHO=1.E5*PL(L)/(RGAS*TL(L))
   TEM=RHO*WMX(L)/(WCONST*FCLD+1.E-20)
   IF(LHX.EQ.LHS) TEM=RHO*WMX(L)/(WMUI*FCLD+1.E=20)
   TEM=TEM*TEM
   IF(TEM.GT.10.) TEM=10.
   CM1=CM0
   IF(BANDF) CM1=CM0*CBF
   IF(LHX.EQ.LHS) CM1=CM0
```

```
CM=CM1*(1.-1./EXP(TEM*TEM))+1.*100.*(PREBAR(L + 1) +
* PRECNVL(L+1)*BYDTsrc)
IF(CM.GT.BYDTsrc) CM=BYDTsrc
PREP(L)=WMX(L)*CM
END IF
C**** FORM CLOUDS ONLY IF RH GT RH00
219 IF(RH1(L).LT.RH00(L)) GO TO 220
C**** COMPUTE THE CONVERGENCE OF AVAILABLE LATENT HEAT
SQ(L)=LHX*QSATL(L)*DQSATDT(TL(L),LHX)*BYSHA
TEM=-LHX*DPDT(L)/PL(L)
QCONV=LHX*AQ(L)-RH(L)*SQ(L)*SHA*PLK(L)*ATH(L)
* -TEM*QSATL(L)*RH(L)
IF(QCONV.LE.0.0.AND.WMX(L).LE.0) GO TO 220
C**** COMPUTE EVAPORATION OF RAIN WATER, ER
RHN=RHF(L)
IF(RHF(L).GT.RH(L)) RHN=RH(L)
```

All climate models treat the laws of physics in the same way, but, since they parameterize phenomena like cloud formation differently, they come up with different results. Also, because the real-world forces influencing the climate are so numerous, different models tend, like medical students, to specialize. GISS's model specializes in the behavior of the atmosphere; other models in the behavior of the oceans; and still others in the behavior of land surfaces and ice sheets.

One rainy November afternoon, I attended a meeting at GISS that brought together members of the institute's modeling team. When I arrived, about twenty men and five women were sitting in battered chairs in a conference room across from Hansen's office. At that particular moment, the institute was performing a series of runs for the U.N. Intergovernmental Panel on Climate Change. The runs were overdue, and apparently the IPCC was getting impatient. Hansen flashed a series of charts on a screen on the wall summarizing some of the results obtained so far.

The obvious difficulty in verifying any particular climate model or climate-model run is the prospective nature of the results. For this reason, models are often run into the past, to see how well they reproduce trends that have already been observed. Hansen told the group that he was pleased with how ModelE had reproduced the aftermath of the eruption of Mount Pinatubo, in the Philippines, which took place in June 1991. Volcanic eruptions release huge quantities of sulfur dioxide—Pinatubo produced some twenty million tons of the gas—which, once in the stratosphere, condenses into tiny sulfate droplets. These droplets, or aerosols, tend to cool the earth by reflecting sunlight back into space. Man-made aerosols, produced by burning coal, oil, and biomass, also reflect sunlight and are a countervailing force to greenhouse warming, albeit one with serious health consequences of its own. (The impact of man-made aerosols is difficult to quantify; without it, however, the earth almost certainly would have warmed even faster than it has.) The cooling effect of aerosols lasts only as long as the droplets remain suspended in the atmosphere. In 1992, following the Pinatubo eruption, global temperatures, which had been rising sharply, fell by half a degree. Then they began to climb again. ModelE had succeeded in simulating this effect to within nine hundredths of a degree. "That's a pretty nice test," Hansen observed laconically.

★   ★   ★

One day, when I was talking to Hansen in his cluttered office, he pulled a pair of photographs out of his briefcase. The first showed a chubby-faced five-year-old girl holding some miniature Christmas-tree lights in front of an even chubbier-faced five-month-old baby. The girl, Hansen told me, was his granddaughter Sophie and the boy was his new grandson, Connor. The caption on the first picture read, "Sophie explains greenhouse warming." The caption on the second photograph, which showed the baby smiling gleefully, read, "Connor gets it."

When modelers talk about what drives the climate, they focus on what they call "forcings." A forcing is any ongoing process or discrete event that alters the energy of the system. Examples of natural forcings include, in addition to volcanic eruptions, periodic shifts in the earth's orbit and changes in the sun's output, like those linked to sunspots. Many climate shifts of the past have no known forcing associated with them; for instance, no one is certain what brought about the so-called Little Ice Age, the cool period that lasted in Europe from around 1500 to 1850. A very large forcing, meanwhile, should produce a commensurately large—and obvious—effect. One GISS scientist put it to me this way: "If the sun went supernova, there's no question that we could model what would happen."

Adding carbon dioxide, or any other greenhouse gas, to the atmosphere by, say, burning fossil fuels or leveling forests is, in the language of climate science, an anthropogenic forcing. Since preindustrial times, the concentration of CO_2 in the atmosphere has risen by roughly a third, from 280 to 378 parts per million. During the same period, the concentration of methane has more than doubled, from .78 to 1.76 parts per million. Scientists measure forcings in terms of watts per square meter, or w/m^2, by which they mean that a certain number of watts have been added (or, in the case of a negative forcing, like aerosols, subtracted) for every single square meter of the earth's surface. The size of the greenhouse forcing is estimated, at this point, to be 2.5 w/m^2. A miniature Christmas light gives off about four tenths of a watt of energy, mostly in the form of heat, so that, in effect (as Sophie supposedly explained to Connor), we have covered the earth with tiny bulbs, six for every square meter. These bulbs are burning twenty-four hours a day, seven days a week, year in and year out.

If greenhouse gases were held constant at today's levels, it is estimated that it would take several decades for the full impact of the forcing that is already in place to be felt. This is because raising the earth's temperature involves not only warming the air and the surface of the land but also melting sea ice, liquefying glaciers, and, most significant, heating the oceans, all processes that require tremendous amounts of energy. (Imagine trying to thaw a gallon of ice cream or warm a pot of water using an Easy-Bake oven.) The delay that is built into the system is, in a certain sense, fortunate. It enables us, with the help of climate models, to foresee what is coming and therefore to prepare for it. But in another sense it is clearly disastrous, because it allows us to keep adding CO_2 to the atmosphere while fobbing the impacts off on our children and grandchildren.

There are two ways to operate a climate model. In the first, which is known as a transient run, greenhouse gases are slowly added to the simulated atmosphere—just as they would be to the real atmosphere—and the model forecasts what the effect of these additions will be at any given moment. In the second, greenhouse gases are added to the atmosphere all at once, and the model is run at these new levels until the climate has fully adjusted to the forcing by reaching a new equilibrium. (Not surprisingly, this is known as an equilibrium run.)

For doubled CO_2, equilibrium runs of the GISS model predict that average global temperatures will rise by 4.9 degrees Fahrenheit. Only about a third of this increase is directly attributable to higher greenhouse gas levels. The rest is a result of indirect effects, like the melting of sea ice, which allows the earth to absorb more heat. The most significant indirect effect is known as the "water-vapor feedback." Since warmer air holds more moisture, higher temperatures are expected to produce an atmosphere containing more water vapor, which is itself a greenhouse gas. GISS's forecast is on the low end of the most recent projections for doubled CO_2; the Hadley Centre model, which is run by the British Met Office, predicts that under these conditions, the eventual temperature rise will be 6.3 degrees Fahrenheit, while Japan's National Institute for Environmental Studies predicts that it will be 7.7 degrees.

In the context of ordinary life, a warming of 4.9, or even of 7.7, degrees may not seem like much to worry about. On the dreary November day I attended the GISS modeling meeting, the temperature in Central Park was fifty-two degrees at seven A.M., and by two P.M. had reached sixty degrees. In the course of a normal summer's day, air temperatures routinely rise by fifteen degrees or more. Average global temperatures, however, have practically nothing to do with ordinary life. This is perhaps best illustrated by the ups and downs of climate history. The so-called Last Glacial Maximum—the point during the most recent glaciation when the ice sheets reached their maximum extent —occurred about twenty thousand years ago. At that time, the Laurentide ice sheet reached deep into what is now the northeastern and midwestern United States, and sea levels were so low that Siberia and Alaska were connected by a land bridge nearly a thousand miles wide. During the Last Glacial Maximum, average global temperatures were only about ten degrees colder than they are today. It is worth noting that the total forcing that ended that ice age is estimated to have been just six and a half watts per square meter.

David Rind is a climate scientist who has worked at GISS since 1978. Rind acts as a troubleshooter for the institute's model, scanning reams of numbers known as diagnostics, trying to catch problems, and he also works with what is known as the GISS Climate Impacts Group. (His office, like Hansen's, is filled with dusty piles of computer printouts.) Although higher temperatures are the most predictable result of increased CO_2, other, second-order consequences—rising sea levels, changes in vegetation, loss of snow cover—are likely to be just as significant. Rind's particular interest is how CO_2 levels will affect water supplies, because, as he put it to me, "you can't have a plastic version of water."

One afternoon, when I was talking to Rind in his office, he mentioned a visit that President George W. Bush's science adviser, John Marburger III, had paid to GISS a few years earlier. "He said, 'We're really interested in adaptation to climate change,' " Rind recalled. "Well, what does 'adaptation' mean?" He rummaged through one of his many file cabinets and finally pulled out a paper that he had published in the *Journal of Geophysical Research* titled "Potential Evapotranspiration and the Likelihood of Future Drought." In much the same way that wind velocity is measured using the Beaufort scale, water availability is measured using the Palmer Drought Severity Index. Different climate models offer very different predictions about future water availability; in the paper, Rind applied the criteria used in the Palmer index to GISS's model and also to a model operated by the National Oceanic and Atmospheric Administration's Geophysical Fluid Dynamics Laboratory. He found that as carbon dioxide levels rose, the world would begin to experience more and more serious water shortages, starting near the equator and then spreading toward the poles. When he applied the index to the GISS model for doubled CO_2, it showed most of the continental United States would be suffering under severe drought conditions. When he applied the index to the GFDL model, the results were even more dire. Rind created two maps to illustrate these findings. Yellow represented a 40 to 60 percent chance of summertime drought, ochre a 60 to 80 percent chance, and brown an 80 to 100 percent chance. In the first map, showing the GISS results, the Northeast was yellow, the Midwest was ochre, and the Rocky Mountain states and California were brown. In the second, showing the GFDL results, brown covered practically the entire country.

"I gave a talk based on these drought indices out in California to water-resource managers," Rind told me. "And they said, 'Well, if that happens, forget it.' There's just no way they could deal with that."

He went on, "Obviously, if you get drought indices like these, there's no adaptation that's possible. But let's say it's not that severe. What adaptation are we talking about? Adaptation in 2020? Adaptation in 2040? Adaptation in 2060? Because the way the models project this, as global warming gets going, once you've adapted to one decade, you're going to have to change everything the next decade.

"We may say that we're more technologically able than earlier societies. But one thing about climate change is it's potentially geopolitically destabilizing. And we're not only more technologically able; we're more technologically able destructively as well. I think it's impossible to predict what will happen. I guess—though I won't be around to see it—I wouldn't be shocked to find out that by 2100 most things were destroyed." He paused. "That's sort of an extreme view."

On the other side of the Hudson River and slightly to the north of GISS, the Lamont-Doherty Earth Observatory occupies what was once a weekend estate in the town of Palisades, New York. The observatory is an outpost of Columbia University, and it houses, among its collections of natural artifacts, the world's largest assembly of ocean-sediment cores—more than thirteen thousand in all. The cores are kept in steel compartments that look like drawers from a filing cabinet, only longer and much skinnier. Some of the cores are chalky, some are clayey, and some are made up almost entirely of gravel. All can be coaxed to yield up—in one way or another—information about past climates.

Peter deMenocal is a paleoclimatologist who has worked at Lamont-Doherty for fifteen years. DeMenocal is an expert on ocean cores, and also on the climate of the Pliocene, which lasted from roughly five million to two million years ago. Around two and a half million years ago, the earth, which had been warm and relatively ice-free, started to cool down until it entered an era—the Pleistocene—of recurring glaciations. DeMenocal has argued that this transition was a key event in human evolution: right around the time that it occurred, at least two types of hominids—one of which would eventually give rise to modern man—branched off from a single ancestral line. Until quite recently, paleoclimatologists like deMenocal rarely bothered with anything much closer to the present day; the current interglacial—the Holocene—was believed to be too stable to warrant much study. In the mid-nineties, though, deMenocal, motivated by a growing concern over global warming— and a concomitant shift in government research funds— decided to look in detail at some Holocene cores. What he learned about the period, as he put it to me when I visited him at Lamont-Doherty, was that it was "less boring than we had thought."

One way to extract climate data from ocean sediments is to examine the remains of what lived or, perhaps more pertinently, what died and was buried there. The oceans are rich with microscopic creatures known as foramini-fera—forams, for short. Forams are tiny, single-celled organisms that construct shells out of calcite. These shells come in a wide range of shapes; viewed under a micro-scope, some look like tiny sand dollars, others like conch shells, and still others like lumps of dough. There are about thirty planktonic species of foraminifera—which is to say, species that live near the top of the sea—and each thrives at a different water temperature, so that by counting a species' prevalence in a given sample it is possible to estimate how warm (or cold) the ocean was at the time the sediment was formed. When deMenocal used this technique to analyze cores that had been collected off the coast of Mauritania, he found that they contained evidence of recurring cool periods; every fifteen hundred years or so, water temperatures would drop for a few centuries before climbing back up again. (The most recent cool period corresponds to the Little Ice Age, which ended about a century and a half ago.) The cores also showed dramatic changes in precipitation. Until about six thousand years ago, northern Africa was relatively wet—dotted with small lakes. Then it became dry, as it is today. DeMenocal traced the shift to periodic variations in the earth's orbit, which, in a generic sense, are the same forces that trigger ice ages. But orbital changes occur gradually, over thousands of years, and northern Africa appears to have switched from wet to dry all of a sudden. Although no one knows exactly how this happened, it seems, like so many climate events, to have been a function of feedbacks—the less rain the continent got, the less vegetation there was to retain water, and so on until, finally, the system just flipped. The process provides yet more evidence of how a very small forcing sustained over time can produce dramatic results.

"We were kind of surprised by what we found," deMenocal told me about his work on the supposedly stable Holocene. "Actually, more than surprised. It was one of these things where, you know, in life you take certain things for granted, like your neighbor's not going to be an ax murderer. And then you discover your neighbor *is* an ax murderer."

Not long after deMenocal began to think about the Holocene, a brief mention of his work on the climate of Africa appeared in a book produced by *National*

Geographic. On the facing page, there was a piece on Harvey Weiss and his work at Tell Leilan. DeMenocal vividly remembers his reaction. "I thought, Holy cow, that's just amazing!" he told me. "It was one of these cases where I lost sleep that night, I just thought it was such a cool idea."

DeMenocal also recalls his subsequent dismay when he went to learn more. "It struck me that they were calling on this climate-change argument, and I wondered how come I didn't know about it," he said. He looked at the *Science* paper in which Weiss had originally laid out his theory. "First of all, I scanned the list of authors and there was no paleoclimatologist on there," deMenocal said. "So then I started reading through the paper and there basically was no paleoclimatology in it." (The main piece of evidence Weiss adduced for a drought was that Tell Leilan had filled with dust.) The more deMenocal thought about it, the more unconvincing he found the data and the more compelling he found the underlying idea. "I just couldn't leave it alone," he told me. In the summer of 1995, he went with Weiss to Syria to visit Tell Leilan. Subsequently, he decided to do his own study to prove—or disprove—Weiss's theory.

Instead of looking in, or near, the ruined city, de-Menocal focused on the Gulf of Oman, a thousand miles downwind. Dust from the Mesopotamian floodplains, just north of Tell Leilan, contains heavy concentrations of the mineral dolomite, and since arid soil produces more wind-borne dust, deMenocal figured that if there had been a drought of any magnitude it would show up in gulf sediments. "In a wet period, you'd be getting none or very, very low amounts of dolomite, and during a dry period you'd be getting a lot," he explained. He and a graduate student named Heidi Cullen developed a highly sensitive test to detect dolomite, and then Cullen assayed, centimeter by centimeter, a sediment core that had been extracted near where the Gulf of Oman meets the Arabian Sea.

"She started going up through the core," deMenocal told me. "It was like nothing, nothing, nothing, nothing, nothing. Then one day, I think it was a Friday afternoon, she goes, 'Oh, my God.' It was really classic." DeMenocal had thought that the dolomite level, if it were elevated at all, would be modestly higher; instead, it went up by 400 percent. Still, he wasn't satisfied. He decided to have the core reanalyzed using a different marker: the ratio of strontium 86 and strontium 87 isotopes. The same spike showed up. When deMenocal had the core carbon-dated, it turned out that the spike lined up exactly with the period of Tell Leilan's abandonment.

Tell Leilan was never an easy place to live. Much like, say, western Kansas today, the Khabur plains received enough annual rainfall—about seventeen inches—to support cereal crops, but not enough to grow much else. "Year-to-year variations were a real threat, and so they obviously needed to have grain storage and to have ways to buffer themselves," deMenocal observed. "One generation would tell the next, 'Look, there are these things that happen that you've got to be prepared for.' And they were good at that. They could manage that. They were there for hundreds of years."

He went on, "The thing they couldn't prepare for was the same thing that we won't prepare for, because in their case they didn't know about it and because in our case the political system can't listen to it. And that is that the climate system has much greater things in store for us than we think."

Shortly before Christmas 2004, Harvey Weiss gave a lunchtime lecture at Yale's Institute for Biospheric Studies. The title was "What Happened in the Holocene," which, as Weiss explained, was an allusion to a famous archae-ology text by V. Gordon Childe, titled *What Happened in History*. The talk brought together archaeological and paleoclimatic records from the Near East over the last ten thousand years.

Weiss, who is sixty years old, has thinning gray hair, wire-rimmed glasses, and an excitable manner. He had prepared for his audience—mostly Yale professors and graduate students—a handout with a time line of Meso-potamian history. Key cultural events appeared in black ink, key climatological ones in red. The two alternated in a rhythmic cycle of disaster and innovation. Around 6200 B.C., a severe global cold snap—red ink—produced ar-idity in the Near East. (The cause of the cold snap is believed to have been a catastrophic flood that emptied an enormous glacial lake—Lake Agassiz—into the North

Atlantic.) Right around the same time—black ink—farming villages in northern Mesopotamia were abandoned, while in central and southern Mesopotamia the art of irrigation was invented. Three thousand years later, there was another cold snap, after which settlements in northern Mesopotamia once again were deserted. The most recent red event, in 2200 B.C., was followed by the dissolution of the Old Kingdom in Egypt, the abandonment of villages in ancient Palestine, and the fall of Akkad. Toward the end of his talk, Weiss, using PowerPoint, displayed some photographs from the excavation at Tell Leilan. One showed the wall of a building—probably intended for administrative offices—that had been under construction when the rain stopped. The wall was made from blocks of basalt topped by rows of mud bricks. The bricks gave out abruptly, as if construction had ceased from one day to the next.

The monochromatic sort of history that most of us grew up with did not allow for events like the drought that destroyed Tell Leilan. Civilizations fell, we were taught, because of wars or barbarian invasions or political unrest. (Another famous text by Childe bears the exemplary title *Man Makes Himself*.) Adding red to the time line points up the deep contingency of the whole enterprise. Civilization goes back, at the most, ten thousand years, even though, evolutionarily speaking, modern man has been around for at least ten times that long. The climate of the Holocene was not boring, but it was dull enough to allow people to sit still. It was only after the immense climatic shifts of the glacial epoch had run their course that agriculture and writing finally emerged.

Nowhere else does the archaeological record go back so far or in such detail as in the Near East. But similar red-and-black chronologies can now be drawn up for many other parts of the world: the Indus Valley, where, some four thousand years ago, the Harappan civilization suffered a decline after a change in monsoon patterns; the Andes, where, fourteen hundred years ago, the Moche abandoned their cities in a period of diminished rainfall; and even the United States, where the arrival of the English colonists on Roanoke Island, in 1587, coincided with a severe regional drought. (By the time English ships returned to resupply the colonists, three years later, not a single one was left.) At the height of the Mayan civilization, population density was five hundred per square mile, higher than it is in most parts of the United States today. Two hundred years later, most Mayan territory had been completely depopulated. You can argue that man through culture creates stability, or you can argue, just as plausibly, that stability is for culture an essential precondition.

After the lecture, I walked with Weiss back to his office, which is near the center of the Yale campus, in the Hall of Graduate Studies. In 2004, Weiss decided to suspend excavation at Tell Leilan. The site lies only fifty miles from the Iraqi border, and, owing to the uncertainties of the war, it seemed like the wrong sort of place to bring graduate students. When I visited, Weiss had just returned from a trip to Damascus, where he had gone to pay the guards who watch over the site when he isn't there. While he was away from his office, its contents had been piled up in a corner by repairmen who had come to fix some pipes. Weiss considered the piles disconsolately, then unlocked a door at the back of the room.

The door led to a second room, much larger than the first. It was set up like a library, except that instead of books the shelves were stacked with hundreds of cardboard boxes. Each box contained fragments of broken pottery from Tell Leilan. Some were painted, others were incised with intricate designs, and still others were barely distinguishable from pebbles. Every fragment had been inscribed with a number, indicating its provenance.

I asked what he thought life in Tell Leilan had been like. Weiss told me that that was a "corny question," so I asked him about the city's abandonment. "Nothing allows you to go beyond the third or fourth year of a drought, and by the fifth or sixth year you're probably gone," he observed. "You've given up hope for the rain, which is exactly what they wrote in 'The Curse of Akkad.' " I said I would like to see something that might have been used in Tell Leilan's last days. Swearing softly, Weiss searched through the rows until he finally found one particular box. It held several potsherds that appeared to have come from identical bowls. They were made from a greenish-colored clay, had been thrown on a wheel, and had no decoration. Intact, the bowls had held about a liter, and Weiss explained that they had been used to mete out rations—probably wheat or barley—to the workers of Tell Leilan. He passed me one of the fragments. I held it in my hand for a moment and tried to imagine the last Akkadian who had touched it. Then I passed it back.

Chapter 6

FLOATING HOUSES

I N FEBRUARY 2003, a series of ads on the theme of
inundation began appearing on Dutch TV. The ads
were sponsored by the Netherlands' Ministry of Transport, Public Works, and Water Management, and they
featured a celebrity weatherman named Peter Timofeeff.
In one commercial, Timofeeff, who looks a bit like
Albert Brooks and a bit like Gene Shalit, sat relaxing
on the shore in a folding chair. "Sea level is rising," he
announced, as waves started creeping up the beach. He
continued to sit and talk even as a boy who had been
building a sand castle abandoned it in panic. At the end
of the ad, Timofeeff, still seated, was immersed in water
up to his waist. In another commercial, Timofeeff was
shown wearing a business suit and standing by a bathtub.
"These are our rivers," he explained, climbing into the
tub and turning on the shower full blast. "The climate is
changing. It will rain more often, and more heavily."
Water filled the tub and spilled over the sides. It dripped
through the floorboards, onto the head of his screeching
wife below. "We should give the water more space and

widen the rivers," he advised, calmly reaching for a towel.

Both the beach chair and the shower ads were part of a public-service campaign titled, somewhat ambiguously, "*Nederland Leeft Met Water*" ("The Netherlands Lives with Water"), which also included radio spots, free tote bags, and newspaper announcements drawn in the form of cartoons. Its tone was consistently lighthearted—other commercials showed Timofeeff trying to start a motorboat in a cow pasture and digging a duck pond in his backyard—either in spite of the fact that, or maybe precisely because, its message for the Dutch was so devastating.

Fully a quarter of the Netherlands lies below sea level, on land wrested from either the North Sea or the Rhine or the Meuse Rivers, or one of the hundreds of natural lakes that once dotted the countryside. Another quarter, while slightly higher, is still low enough that, in the natural course of events, it would regularly be flooded. What has made this arrangement possible is the world's most sophisticated water management system, which, according to government figures, comprises 150 miles of dunes, 260 miles of sea dikes, 850 miles of river dikes, 610 miles of lake dikes, and 8,000 miles of canal dikes, not to mention countless pumps, holding ponds, and windmills.

Historically, whenever flooding has occurred, the Dutch response has been either to reinforce the dikes or to add new ones. In 1916, for example, after the defenses gave out along an inlet of the North Sea known as the Zuiderzee, the Dutch dammed up the Zuiderzee, creating an artificial lake as large as Los Angeles. In 1953, storms overwhelmed the dikes in the province of Zeeland, killing 1,835 people. Immediately afterward, the government embarked on a massive, five-and-a-half-billion-dollar construction project known as the Delta Works. (The last phase of the project, the Maeslant barrier, which was finally completed in 1997, is supposed to protect Rotterdam from storm surges with the aid of two moveable arms, each the size of a skyscraper.) People in Holland like to joke, although they are not really joking, "God made the world, but the Dutch made the Netherlands."

"The Netherlands Lives with Water" signals the end of this five-hundred-year project. Looking ahead, the same engineers who built the Maeslant barrier have determined that even such monumental public works projects are no longer adequate. From now on, instead of reclaiming land from the water, the Dutch, they have decided, are going to have to start giving back.

The Nieuwe Merwede looks like a river, but is actually a canal, dug in the 1870s in the delta of the Rhine and Meuse Rivers. It runs on a winding course west from the city of Werkendam until it meets up with another man-made river to form what is known as the Hollandsch Diep, which, in the confusion of the delta, splits again, and eventually empties into the North Sea.

On the north side of the canal, in a pocket-sized national park called Biesbosch, is a nature center, which, at the time that I visited, was running an exhibit on climate change. By way of decoration, large black umbrellas had been hung from the ceiling, and in the background, the soundtrack of a church bell—the traditional Dutch flood warning—tolled periodically. One kid-friendly display allowed visitors to turn a crank and, in effect, drown the countryside. By 2100, the display showed, the Nieuwe Merwede will be running, at peak flows, several feet above the top of the local dikes.

There are several reasons why global warming produces flooding. The first has to do simply with the physics of liquids. As water warms, it expands. In a small body of water, the effect is small; in a big body, it's commensurately larger. Most of the sea level rise predicted for the next hundred years—a total of up to three feet—is purely a function of thermal expansion. (Even if greenhouse gas levels are eventually stabilized, sea levels will continue to rise for several centuries, owing to the oceans' thermal inertia.)

Meanwhile, a warming Earth means changing precipitation patterns. Just as some regions, like the American Midwest, are predicted to suffer from drought, others will experience more—or at least more intense—rainfall. The effect is likely to be particularly punishing in some of the most densely populated regions on Earth, including the Mississippi Delta, the Ganges Delta, and the Thames basin. A study commissioned a few years ago by the British government concluded that under certain conditions, floods of a magnitude now expected no more often than once a century could, by 2080, be occurring in England once every three years. (As it happened, the very week I was in the Netherlands, thirteen people were killed by exceptionally heavy winter storms in Britain and Scandinavia.)

At the Biesbosch nature center, I met up with a water-ministry official named Eelke Turkstra. Turkstra runs a program called *Ruimte voor de Rivier* (Room for the River), and these days his job consists not in building dikes, but in dismantling them. He explained to me that the Dutch were already seeing more rainfall than they used to. Where once the water ministry had planned on peak flows in the Rhine of no more than fifteen thousand cubic meters per second, recently it had been forced to raise that to sixteen thousand cubic meters per second and was already anticipating having to deal with eighteen thousand cubic meters per second. Rising sea levels, meanwhile, were likely to further compound the problem by impeding the flow of the river to the ocean.

"We think in the run of this century, sea levels can rise by sixty centimeters," or just under two feet, Turkstra told me. "When that happens—we're sure that it *will* happen—that makes things very complicated."

From the nature center, we took a car ferry across the Nieuwe Merwede. The area we were driving through was made up entirely of "polders"—land that has been laboriously reclaimed from the water. The polders were shaped like ice trays, with sloping sides and perfectly flat fields along the bottom. Every once in a while, there was a sturdy-looking farmhouse. The whole scene—the level fields, the thatched barns, even the gray clouds sitting on the horizon—could have been borrowed from a painting by Hobbema. All this land, Turkstra said, was destined for inundation. The plan of Room for the River was to buy out the farmers who were living in the polders, and then lower the surrounding dikes. By selectively abandoning rural areas like this one, the water ministry hoped to be able to protect population centers like the nearby city of Gorinchem. The price tag for the project we were looking at had been set at $390 million. Similar projects were under way in other parts of the Netherlands, and still others were in the design phase. Some of the designs had provoked angry, ongoing protests. Surrendering land that people have been living on for decades, in some cases centuries, was, Turkstra acknowledged, bound to cause political problems, but that was precisely the reason that it was important to get started immediately.

"Some people don't get it," he told me as we zipped along. "They think this project is stupid. But I think it's stupid to continue the old way."

When climatologists discuss the hazards of rising greenhouse gas levels, they use the phrase "dangerous anthropogenic interference" or, for short, DAI. The term does not refer to any disaster in particular, although there are, it is generally agreed, a number of scenarios that would fit the bill—climate change dramatic enough to destroy entire ecosystems, for instance, or cause mass extinction or disrupt the world's food supply. The disintegration of one of the planet's remaining ice sheets is often held up as the exemplary catastrophe. The West Antarctic ice sheet is, at this point, the world's only marine ice sheet, meaning that it rests on land that is below sea level. For this reason it is considered particularly vulnerable to collapse. Were the West Antarctic or the Greenland ice sheet to be destroyed, sea levels around the world would rise by at least fifteen feet. Were both ice sheets to disintegrate, global sea levels would rise by thirty-five feet. It could take centuries for either of the ice sheets to disappear entirely, but once disintegration got under way it would start to feed on itself, most likely becoming irreversible. Other catastrophes have similar built-in delays, which follow from the tremendous inertia of the climate system. DAI is therefore understood to refer not to the end of the process—the moment when disaster actually arrives—but to the beginning of it: the point at which its arrival becomes unavoidable.

Exactly what forcing or temperature or level of CO_2 represents DAI is a question of the utmost significance and one that cannot at this point be answered. Policy studies often take 500 parts per million of CO_2—roughly double preindustrial levels—as the threshold. But this figure has at least as much to do with what appears to be a socially feasible goal as with what has been scientifically demonstrated.

In the last decade, a great deal has been discovered about how the climate functions, both through measurements made in real time and through reconstructions of the paleoclimatic record. Just about everything that has been learned—from the observed acceleration of the ice sheets

to the inferred history of the thermohaline circulation—
has tended to push the level of DAI downward. Many
climate scientists now believe that 450 parts per million of
CO_2 represents a more objective estimate of danger, while
others argue that the threshold is 400 parts per million or
even lower.

Probably the most significant of the recent discoveries
was made in Antarctica, at a research base known as the
Vostok station. Between 1990 and 1998, an 11,775-foot-
long ice core was drilled there. Since less snow falls in
Antarctica than in Greenland, the layers in an Antarctic
core are thinner and the climate information contained in
them is less detailed. However, they go back much farther.
The Vostok core, which is now stored in pieces in Denver,
Grenoble, and on Antarctica, contains a continuous climate record stretching back four full glacial cycles. (As is
the case with Greenland cores, temperatures can be ascer-
tained by measuring the isotopic composition of the ice,
and the makeup of the atmosphere determined by analyzing tiny bubbles of trapped air.)

What the Vostok record shows is that the planet is
already nearly as warm as it has been at any point in the last
420,000 years. A possible consequence of even a four- or
five-degree temperature rise—on the low end of projections for the end of this century—is that the world will
enter a completely new climate regime, one with which
modern humans have no prior experience. When it comes
to carbon dioxide, meanwhile, the evidence is even more
striking. The Vostok record demonstrates that, at 378 parts

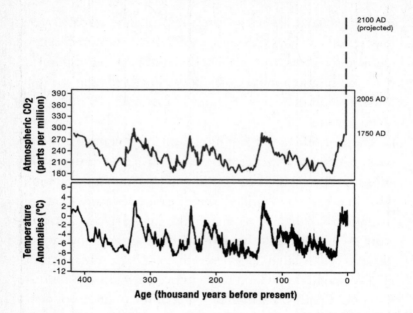

The record from the Vostok core shows that CO_2 levels and temperatures have varied in tandem. Current CO_2 levels are unprecedented in the last 420,000 years. Credit: J.R. Petit et al, Nature, vol. 399 (1999).

per million, current CO_2 levels are unprecedented in recent geological history. (The previous high, of 299 parts per million, was reached around 325,000 years ago). It is believed that the last time carbon dioxide levels were comparable to today's was three and a half million years ago, during what is known as the mid-Pliocene warm period, and it is likely that they have not been much higher since the Eocene, some fifty million years ago. In the Eocene, crocodiles roamed Colorado and sea levels were nearly three hundred feet higher than they are today. A scientist with the National Oceanic and Atmospheric Administration (NOAA) put it to me—only half-jokingly—this way: "It's true that we've had higher CO_2 levels before. But, then, of course, we also had dinosaurs."

The town of Maasbommel is situated about fifty miles east of Biesbosch. It lies on the banks of the River Meuse and is a popular holiday destination; every summer it fills with tourists who have come to go boating or to camp out. Thanks to the risk of flooding, building is restricted along the river, but a few years ago one of the Netherlands' largest construction firms, Dura Vermeer, received permission to turn a former RV park on the banks of the Meuse into a development of "amphibious homes."

The first of the amphibious homes were completed in the fall of 2004, and on a dull winter's day a few months afterward, I went to take a look at them. On my way, I stopped off at Dura Vermeer's headquarters to meet with the company's environmental director, Chris Zevenbergen. In his office, Zevenbergen played for me an animated video on the future of the Netherlands; it showed large chunks of the country gradually being swallowed up by water. It was lunchtime, and after a while his secretary came around carrying a tray of sandwiches and a large pitcher of milk. Zevenbergen explained that Dura Vermeer was also working to construct buoyant roads and floating greenhouses. While each of these projects represents a somewhat different engineering challenge, they have a common goal, which is to allow people to continue to inhabit areas that, periodically at least, will be inundated. "There is a flood market emerging," Zevenbergen told me.

From the company's headquarters, it was about an hour's drive to Maasbommel. By the time I arrived, the sun was starting to sink, and in the afternoon light, the Meuse was glowing silver.

The amphibious homes all look alike. They are tall and narrow, with flat sides and curved metal roofs, so that standing next to one another they resemble a row of toasters. Each one is moored to a metal pole and sits on a set of hollow concrete pontoons. Assuming that all goes according to plan, when the Meuse floods, the homes will bob up and then, when the water recedes, they will gently be deposited back on land. At the point that I visited, a half a dozen families were occupying their amphibious houses. Anna van der Molen, a nurse and mother of four, gave me a tour of hers. She was enthusiastic about life on the river. "Not one day is the same," she told me. In the future, she said, she expected that people all over the world would live in floating houses, since, as she put it, "the water is coming up, and we have to live with it, not fight it—it's just not possible."

Chapter 7

BUSINESS AS USUAL

I N CLIMATE-SCIENCE CIRCLES, a future in which current emissions trends continue, unchecked, is known as "business as usual," or BAU. About five years ago, Robert Socolow, a professor of engineering at Princeton, began to think about BAU and what it implied for the fate of mankind. At that point, Socolow had recently become codirector of the Carbon Mitigation Initiative, a project funded by BP and Ford, but he still considered himself an outsider to the field of climate science. Talking to insiders, he was struck by the degree of their alarm. "I've been involved in a number of fields where there's a lay opinion and a scientific opinion," he told me when I went to visit him at his office shortly after returning from the Netherlands. "And, in most of the cases, it's the lay community that is more exercised, more anxious. If you take an extreme example, it would be nuclear power, where most of the people who work in nuclear science are relatively relaxed about very low levels of radiation. But, in the climate case, the experts— the people who work with the climate models every day, the people who do ice cores—they are *more* concerned. They're going out of their way to say, 'Wake up! This is not a good thing to be doing.' "

Socolow, who is sixty-seven, is a trim man with wire-rimmed glasses and gray, vaguely Einsteinian hair. Although by training he is a theoretical physicist—he did his doctoral research on quarks—he has spent most of his career working on problems of a more human scale, like how to prevent nuclear proliferation or construct buildings that don't leak heat. In the 1970s, Socolow helped design an energy-efficient housing development in Twin Rivers, New Jersey. At another point, he developed a system—never commercially viable—to provide air-conditioning in the summer using ice created in the winter. When Socolow became codirector of the Carbon Mitigation Initiative, he decided that the first thing he needed to do was get a handle on the scale of the carbon problem. He found that the existing literature on the subject offered almost too much information. In addition to BAU, a dozen or so alternative scenarios, known by code names like A1 and B1, had been devised; these all tended to jumble together in his mind, like so many Scrabble tiles. "I'm pretty quantitative, but I could not remember these graphs from one day to the next," he recalled. He decided to try to streamline the problem, mainly so that he could understand it.

Here in the United States, most of us begin generating CO_2 as soon as we get out of bed. Seventy percent of our electricity is generated by burning fossil fuels—a little more than 50 percent from burning coal and another 17 percent from natural gas—so that to turn on the lights is, indirectly at least, to pump carbon dioxide into the atmosphere. Making a pot of coffee, either on an electric or a gas range, adds more emissions, as does taking a hot shower, watching the morning news on TV, and driving to work. Exactly how much CO_2 any particular action produces depends on a variety of factors. Though all fossil fuels produce carbon dioxide as an inevitable product of combustion, some fuels, most notably coal, give off more than others for each unit of power generated. A kilowatt-hour of electricity delivered from a coal-fired plant will produce slightly more than half a pound of carbon, while if the power is originating from a plant that runs on natural gas, it will produce roughly half that amount. (When measuring CO_2, it is customary to count not the full weight of the gas, but just the weight of the carbon—to convert back, multiply by 3.7.) Every gallon of gasoline that is consumed produces about five pounds of carbon, meaning that in the course of a forty-mile commute, a vehicle like a Ford Explorer or a GM Yukon throws about a dozen pounds of carbon into the air. On average, every single person in America generates twelve thousand pounds of carbon per year. (If you would like to figure out your own annual contribution to green-house warming, go to the Environmental Protection Agency's Web site and plug various facts about your lifestyle—what kind of car you drive, how much of your trash you recycle, and so on—into the "personal emissions calculator" provided there.) The largest single source of carbon emissions in the United States is electricity production, at 39 percent, followed by transportation, at 32 percent. In a country like France, where three quarters of the power is produced by nuclear plants, this ratio is very different, and it's different again in countries like Bhutan, where many people don't even have access to electricity and where they burn wood and animal waste to cook and heat their homes.

In the future, the growth of carbon emissions is likely to be determined by several forces. One is the rate of population growth; estimates of how many people will be living on the planet in 2050 range from a low of 7.4 billion to a high of 10.6 billion. Another is economic growth. A third factor is the rate at which new technologies are adopted. Particularly in the developing world, the demand for electricity is increasing rapidly; in China, for example, electricity consumption is expected to more than double by 2025. If developing nations satisfy this demand by adopting the latest, most energy-efficient technologies, then emissions will grow at one rate. (This possibility is sometimes referred to as "leapfrogging," since it would require developing countries to "leapfrog" ahead of industrialized nations.) If they satisfy demand by deploying less efficient—but often cheaper—technologies, emissions will increase at a much faster rate.

"Business as usual" refers to a whole range of projections, all of which take as their primary assumption that emissions will continue to grow without regard to the climate. In 2005, global emissions amounted to roughly 7 billion metric tons of carbon. Under a midrange BAU projection, they will grow to 10.5 billion metric tons a year by 2029, and 14 billion tons a year by 2054. Under this same projection, CO_2 levels in the atmosphere will reach 500 parts per million by the middle of the century, and if things continue on the same trajectory, CO_2 will reach 750 parts per million, or roughly three times preindustrial levels, by the year 2100.

Looking at these figures, Socolow reached a couple of conclusions right away. The first was that to avoid exceeding CO_2 concentrations of 500 parts per million, immediate action would be needed. The second was that to meet this target, emissions growth would have to be held essentially to zero. Stabilizing CO_2 emissions would be such an enormous undertaking that Socolow decided to break the problem down into more manageable blocks, which he called "stabilization wedges." For simplicity's sake, he defined a stabilization wedge as a step that would be sufficient to prevent a billion metric tons of carbon per year from being emitted by 2054. Since annual carbon emissions now amount to 7 billion metric tons, and in fifty years are expected to reach 14 billion metric tons, seven wedges would be needed to hold emissions constant at today's level. With the help of a Princeton colleague, Stephen Pacala, Socolow eventually came up with fifteen different wedges— theoretically, at least, eight more than would be necessary. In August 2004, Socolow and Pacala published their findings in a paper in *Science* that received a great deal of attention. The paper was at once upbeat—"Humanity already possesses the fundamental scientific, technical, and industrial know-how to solve the carbon and climate problem for the next half-century," it declared—and deeply sobering. "There is no easy wedge" is how Socolow put it to me.

Consider wedge No. 11. This is the photovoltaic, or solar power, wedge—probably the most appealing of all the alternatives, at least in the abstract. Photovoltaic cells, which have been around for more than fifty years, are already in use in all sorts of small-scale applications and in some larger ones where the cost of connecting to the electrical grid is prohibitively high. The technology, once installed, is completely emissions-free, producing no waste products, not even water. For the purpose of their calculations, Socolow and Pacala assumed that a one-thousand-megawatt coal-fired power plant would produce about 1.5 million tons of carbon a year. (Today's coal plants actually produce some 2 million tons of carbon a year, but in the future, plants are expected to become more efficient.) To reduce emissions by a billion metric tons a year, enough solar cells would therefore have to be installed to obviate the need for nearly seven hundred thousand-megawatt coal plants. Since sunshine is not constant—it is interrupted by nightfall and by clouds—some two million megawatts of capacity would be needed. This, it

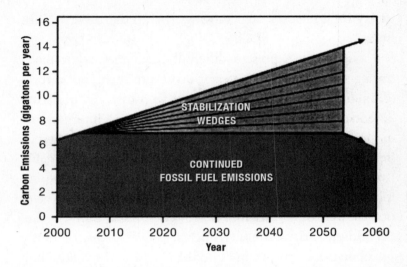

One "wedge" would prevent a billion tons of carbon a year from being emitted by 2054. Credit: S. Pacala and R. Socolow, Science, *vol. 305 (2004).*

turns out, would require PV arrays covering a surface area of five million acres—approximately the size of Connecticut.

Wedge No. 10 is wind electricity. Again, the technology has the advantage of being both safe and emissions-free. A large turbine can generate two megawatts of power, but since the wind, like sunlight, is intermittent, to get a wedge out of wind power would require at least a million two-megawatt turbines. Wind turbines are generally installed either offshore, or on hilltops or windy plains. When they are installed on land, the area around them can be used for other purposes, such as farming, but a million turbines would effectively "occupy" thirty million acres, an area roughly the size of New York state.

Other wedges present different challenges, some technical, some social. Nuclear power produces no carbon dioxide, but it generates radioactive waste, with all the attendant difficulties of storage, disposal, and international policing. More than forty years after the first commercial reactors went online, the United States has been unable to solve its nuclear waste problems, and several power plant operators have sued the federal government over its failure to construct a long-term waste storage site. Worldwide, there are 441 nuclear power plants currently in operation; one wedge could be achieved by doubling their capacity. There is also one heating and lighting wedge, which would result from cutting energy use in residential and commercial buildings by a quarter, and two automobile wedges. The first auto wedge would require that every car in the world be driven half as much as it is today, the second that it be twice as efficient. (Since the late 1980s, the fuel efficiency of passenger vehicles in the United States has actually declined, by more than 5 percent.)

Another possible option is a technology known as "carbon capture and storage," or CCS. As the name suggests, with CCS carbon dioxide is "captured" at the source—presumably a large emitter—and then injected at very high pressure into geological formations, such as depleted oil fields, underground. (At such pressure, CO_2 becomes "supercritical," a phase in which it is not quite a liquid and not quite a gas.) One wedge in Socolow's plan comes from "capturing" CO_2 from power plants, another from capturing it from synthetic-fuel manufacturers. The basic techniques of CCS are currently employed to increase production from oil and natural gas wells. However, at this point, there are no synthetic-fuel or power plants using the process. Nor does anyone know for certain how long CO_2 injected underground will remain there. The world's longest-running CCS effort, maintained by the Norwegian oil company Statoil at a natural gas field in the North Sea, has been operational only for about a decade. A wedge of CCS would require thirty-five hundred projects on the scale of Statoil's.

In a world like today's, where there is, for the most part, no direct cost to emitting CO_2, none of Socolow's wedges are apt to be implemented; this is, of course, why they represent a departure from "business as usual." To alter the economics against carbon requires government intervention. Countries could set a strict limit on CO_2, and then let emitters buy and sell carbon "credits." (In the United States, this same basic strategy has been used successfully with sulfur dioxide in order to curb acid rain.) Another alternative is to levy a tax on carbon. Both of these options have been extensively studied by economists; using their work, Socolow estimates that the cost of emitting carbon would have to rise to around a hundred dollars a ton to provide a sufficient incentive to adopt many of the options he has proposed. Assuming that the cost were passed on to consumers, a hundred dollars a ton would raise the price of a kilowatt-hour of coal-generated electricity by about two cents, which would add roughly fifteen dollars a month to the average American family's electricity bill.

All of Socolow's calculations are based on the notion—clearly hypothetical—that steps to stabilize emissions will be taken immediately, or at least within the next few years. This assumption is key not only because we are constantly pumping more CO_2 into the atmosphere but also because we are constantly building infrastructure that, in effect, guarantees that that much additional CO_2 will be released in the future. In the United States, the average new car gets about twenty miles to the gallon; if it is driven a hundred thousand miles, it will produce more than eleven metric tons of carbon. A thousand-megawatt coal plant built today, meanwhile, is likely to last fifty years and to emit some hundred million tons of carbon during its life. The overriding message of Socolow's wedges is that the longer we wait—and the more infrastructure we build without regard to its impact on emissions—the more daunting the task of keeping CO_2 levels below 500 parts per million will become.

Indeed, even if we were to hold emissions steady for the next half century, Socolow's graphs show that much steeper cuts would be needed in the following half century to keep CO_2 concentrations from exceeding that level. Carbon dioxide is a persistent gas; it lasts for about a century. Thus, while it is possible to increase CO_2 concentrations relatively quickly, the opposite is not the case. (The effect might be compared to driving a car equipped with an accelerator but no brakes.) After a while, I asked Socolow whether he thought that stabilizing emissions was a politically practical goal. He frowned.

"I'm always being asked, 'What can you say about the practicability of various targets?'" he told me. "I really think that's the wrong question. These things can all be done.

"What kind of issue is like this that we faced in the past?" he continued. "I think it's the kind of issue where something looked extremely difficult, and not worth it, and then people changed their minds. Take child labor. We decided we would not have child labor and goods would become more expensive. It's a changed preference system. Slavery also had some of those characteristics a hundred and fifty years ago. Some people thought it was wrong, and they made their arguments, and they didn't carry the day. And then something happened and all of a sudden it was wrong and we didn't do it anymore. And there were social costs to that. I suppose cotton was more expensive. We said, 'That's the trade-off; we don't want to do this anymore.' So we may look at this and say, 'We are tampering with the earth.' The earth is a twitchy system. It's clear from the record that it does things that we don't fully understand. And we're not going to understand them in the time period we have to make these decisions. We just know they're there. We may say, 'We just don't want to do this to ourselves.' If it's a problem like that, then asking whether it's practical or not is really not going to help very much. Whether it's practical depends on how much we give a damn."

Marty Hoffert is a professor of physics at New York University. He is big and bearish, with a wide face and silvery hair. Hoffert got his undergraduate degree in aeronautical engineering, and one of his first jobs, in the mid-1960s, was helping to develop the United States's antiballistic-missile system. During the week, Hoffert worked at a lab in New York, and sometimes he would go down to Washington to meet with Pentagon officials. Over the weekend, on occasion, he would travel back to Washington to protest Pentagon policy. Eventually, he decided that he wanted to work on something, in his words, "more productive." In this way, he became involved in climate research. He calls himself a "technological optimist," and a lot of his ideas about electric power have a wouldn't-it-be-cool, Buck Rogers sound to them. On other topics, though, Hoffert is a killjoy.

"We have to face the quantitative nature of the challenge," he told me one day over lunch at the NYU faculty club. "Right now, we're going to just burn everything up; we're going to heat the atmosphere to the temperature it was in the Cretaceous, when there were crocodiles at the poles. And then everything will collapse."

Hoffert is primarily interested in finding new, carbon-free ways to generate energy. Currently, the technology that he is pushing is space-based solar power, or SSP. In theory, at least, SSP involves launching into space satellites equipped with massive photovoltaic arrays. Once a satellite is in orbit, the array would unfold or, according to some plans, inflate. SSP has two important advantages over conventional, land-based solar power. In the first place, there is more sunlight in space—roughly eight times as much, per unit of area—and, in the second, this sunlight is constant: satellites are not affected by clouds or by nightfall. The obstacles, meanwhile, are several. No full-scale test of SSP has ever been conducted. (In the 1970s, NASA studied the idea of sending a photovoltaic array the size of Manhattan into space, but the project never, as it were, got off the ground.) Then, there is the expense of launching satellites. Finally, once the arrays are up, there is the difficulty of getting the energy down. Hoffert imagines solving this last problem by using microwave beams of the sort used by cell phone towers, only much more tightly focused. He believes, as he put it to me, that SSP has a great deal of "long-term promise"; however, he is quick to point out that he is open to other ideas, like putting solar collectors on the moon, or using superconducting wires to transmit electricity with minimal energy loss, or generating wind power using turbines suspended in the jet stream. The important thing, he says, is not *which* new technology will work but simply that *some* new technology be found: "There's an argument that our civilization can continue to exist with the present number of people and the present kind of high technology through conservation. I see that argument as similar to a man being locked in a sealed room with a limited amount of oxygen. And if he breathes more slowly, he'll be able to live longer, but what he really needs is to get out of the room. And I want to get out of the room." A few years ago, Hoffert published an influential paper in *Science* in which he argued that holding CO_2 levels below 500 parts per million would require a "Herculean" effort and probably could be accomplished only through "revolutionary" changes in energy production.

"The idea that we already possess the 'scientific, technical, and industrial know-how to solve the carbon problem' is true in the sense that, in 1939, the technical and scientific expertise to build nuclear weapons existed," he told me, quoting Socolow. "But it took the Manhattan Project to make it so."

Hoffert's primary disagreement with Socolow, which both men took pains to point out to me and also took pains to try to minimize, is over the future trajectory of CO_2 emissions. For the past several decades, as the world has turned increasingly from coal to oil, natural gas, and nuclear power, emissions of CO_2 per unit of energy have declined, a process known as "decarbonization." This has slowed the growth of emissions relative to the growth of the global economy; without it, CO_2 levels today would be significantly higher.

In the "business as usual" scenario that Socolow uses, it is assumed that decarbonization will continue. To assume this, however, is to overlook several emerging trends. Most of the growth in energy usage in the next few decades is due to occur in places like China and India, where supplies of coal far exceed those of oil or natural gas. (China, which is adding new coal-fired generating capacity at the rate of more than a gigawatt a month, is expected to overtake the United States as the world's largest carbon emitter around 2025.) Meanwhile, global production of oil and gas is expected to start to decline—according to some experts in twenty or thirty years, and to others by the end of this decade. Hoffert predicts that the world will start to "recarbonize," a development that would make the task of stabilizing carbon dioxide that much more difficult. By his accounting, recarbonization will mean that as many as twelve wedges will be needed simply to keep CO_2 emissions on the same upward trajectory they're on now. (Socolow readily acknowledges that there are plausible scenarios that would push up the number of wedges needed.) Hoffert told me that he thought the federal government should be budgeting between ten and twenty billion dollars a year for primary research into new energy sources. For comparison's sake, he pointed out that the "Star Wars" missile-defense program, which still hasn't yielded a workable system, has already cost the government nearly a hundred billion dollars.

A commonly heard argument against acting to curb

global warming is that the options now available are inadequate. To his dismay, Hoffert often finds his ideas being cited in support of this argument, with which, he says, he vigorously disagrees. "I want to make it very clear," he told me at one point. "We have to start working immediately to implement those elements that we know how to implement *and* we need to start implementing these longer-term programs. Those are not opposing ideas."

"Let me say this," he said at another point. "I'm not sure we can solve the problem. I hope we can. I think we have a shot. I mean, it may be that we're not going to solve global warming, the earth is going to become an ecological disaster, and, you know, somebody will visit in a few hundred million years and find there were some intelligent beings who lived here for a while, but they just couldn't handle the transition from being hunter-gatherers to high technology. It's certainly possible. Carl Sagan had an equation—the Drake equation—for how many intelligent species there are in the galaxy. He figured it out by saying, How many stars are there, how many planets are there around these stars, what's the probability that life will evolve on a planet, what's the probability if you have life evolve of having intelligent species evolve, and, once that happens, what's the average lifetime of a technological civilization? And that last one is the most sensitive number. If the average lifetime is about a hundred years, then probably, in the whole galaxy of four hundred billion stars, there are only a few that have intelligent civilizations.

If the lifetime is several million years, then the galaxy is teeming with intelligent life. It's sort of interesting to look at it that way. And we don't know. We could go either way."

Chapter 8

THE DAY AFTER KYOTO

WHEN THE KYOTO PROTOCOL went into effect, on February 16, 2005, the event was seen as a cause for celebration in many cities around the world. The mayor of Bonn hosted a reception in the Rathaus; Oxford University held an "Entry into Force" banquet; and in Hong Kong there was a Kyoto prayer meeting. As it happened, that day, an exceptionally warm one in Washington, D.C., I went to speak to the Under Secretary of State for Democracy and Global Affairs, Paula Dobriansky.

Dobriansky is a slight woman with shoulder-length brown hair and a vaguely anxious manner. Among her duties is explaining the Bush administration's position on global warming to the rest of the world, a task that, on the occasion of Kyoto's entry into force, seemed peculiarly unenviable. The United States is by far the world's largest emitter of greenhouse gases in aggregate—it produces nearly a quarter of the world's total—and on a per capita basis is rivaled only by a handful of nations, like Qatar. Yet the United States is one of only two industrialized nations that have rejected the Kyoto Protocol, and, with it, mandatory cuts in emissions. (The other outlier is Australia.) Two of Dobriansky's assistants accompanied me into her office. We all took seats in a circle.

Dobriansky began by assuring me that despite how it might appear, the Bush administration took the issue of climate change "very seriously." She went on, "Also let me just add, because in terms of taking it seriously, not only stating to you that we take it seriously, we have engaged many countries in initiatives and efforts, whether they are bilateral initiatives—we have some fourteen bilateral initiatives—and in addition we have put together some multilateral initiatives. So we view this as a serious issue." I asked her how, then, the administration justified its position on Kyoto to its allies. "We have a common goal and objective," she replied. "Where we differ is on what approach we believe is and can be the most effective." A few moments later, she added, as if expanding on this statement: "The bottom line here is, in grappling with a serious issue, we believe we have a common goal and objective, but that we can take different approaches."

The remainder of our brief conversation followed much the same lines. At one point, I asked the undersecretary if there were any circumstances under which the administration would accede to mandatory caps on emissions. "Our approach has been predicated on: we act, we learn, we act again," she said. In response to a question about how urgent the problem of stabilizing emissions was, she replied, "We act, we learn, we act again," and in response to a question about what would constitute a "dangerous" level of CO_2 in the atmosphere, she said, "Forgive me, I'm going to repeat myself: we act, we learn, we act again." Dobriansky told me twice that the administration's approach to global warming encompassed both "near-term actions and long-term actions" and three times that it saw economic growth as "the solution, not the problem." I had been instructed that Dobriansky could spare no more than twenty minutes. According to my digital recorder, after fifteen minutes and thirty-five seconds one of her assistants announced that it was time to wrap things up. As I was getting ready to leave, I asked Dobriansky if there was anything more she wanted to say.

"I'd say this to you," she replied. "We see this as a serious issue. We have vigorously and robustly put forth a climate change policy to address these issues, and we will continue to work with other countries to address the issue of climate change. Basically and fundamentally we have a common goal and objective, but we are pursuing different approaches."

On paper at least, the United States, along with the rest of the world, has been committed to addressing global warming for nearly fifteen years. In June 1992, the United Nations held the so-called Earth Summit in Rio de Janeiro, which was attended by more than twenty thousand people. Representatives from virtually every country on the globe met there to discuss and ultimately endorse the U.N. Framework Convention on Climate Change. One of the earliest signatories was President George H. W.

Bush, who, while in Rio, called on world leaders to translate "the words spoken here into concrete action to protect the planet." Three months later, Bush submitted the Framework Convention to the U.S. Senate, which approved it by unanimous consent.

In the English version, the Framework Convention runs to thirty-three pages. It starts with vague statements of principle ("Acknowledging that change in the Earth's climate and its adverse effects are a common concern of humankind . . ."; "Concerned that human activities have been substantially increasing the atmospheric concentrations of greenhouse gases . . .") and works its way through a long list of definitions ("'Climate change' means a change of climate which is attributed directly or indirectly to human activity"; "'Climate system' means the totality of the atmosphere, hydrosphere, biosphere and geosphere and their interactions") before finally arriving at its objective. This is: the "stabilization of greenhouse gas concentrations in the atmosphere at a level that would prevent dangerous anthropogenic interference with the climate system."

Every country that signed on to the Framework Convention accepted the same goal—avoiding DAI. But not every country accepted the same obligations. The treaty distinguished between industrialized nations, which, in U.N.-speak, became known as the Annex 1 countries, and basically everyone else. While the latter group agreed to take steps to "mitigate" climate change, the former agreed to reduce its greenhouse gas emissions. (In diplomatic terms, this arrangement followed the principle of "common but differentiated responsibilities.") Article 4, paragraph 2, subparagraph b of the Framework Convention spelled out what compliance meant; it instructed Annex 1 countries, which include the United States, Canada, Japan, and the nations of Europe and the erstwhile Eastern bloc—to "aim" to return their emissions to 1990 levels.

As it turned out, submitting the Framework Convention to the Senate was one of George Bush Senior's last acts as president. Bill Clinton reaffirmed U.S. support of the convention, announcing, shortly after taking office, on Earth Day 1993, that the nation was committed to reducing its greenhouse gas emissions to 1990 levels by the year 2000. "Unless we act now," he said, "we face a future in which the sun may scorch us, not warm us; where the change of season may take on a dreadful new meaning; and where our children's children will inherit a planet far less hospitable than the world in which we came of age."

Yet even as Clinton was reasserting the nation's commitment, emissions in the United States and indeed around the globe were continuing to rise. By 1995, pretty much the only countries that were making any progress toward compliance were former members of the Soviet bloc, and this was because their economies were in free fall. Meanwhile, as emissions continued to go up, what had initially seemed a rather modest goal—returning to 1990 levels—started to look more and more ambitious. Several rounds of often bitter negotiations followed—in Berlin in March

1995, in Geneva in July 1996, and, finally, in Kyoto in December 1997.

Technically, the agreement that emerged from the Kyoto session is simply an addendum to the Framework Convention. (Its full title is the Kyoto Protocol to the United Nations Framework Convention on Climate Change.) The protocol has the same goal as the convention—avoiding DAI—and hews to the same principle of "common but differentiated responsibilities." But for vague exhortations, like "aim," the protocol substitutes mandatory commitments. Exactly what these commitments are varies slightly from country to country, based on a combination of historical and political factors. The nations of the European Union, for example, are supposed to reduce their greenhouse gas emissions 8 percent below 1990 levels and to do so by 2012, the year that the protocol lapses. The United States, meanwhile, has a target of 7 percent below 1990 levels, and Japan has a target of 6 percent below. The treaty covers five greenhouse gases in addition to CO_2—methane, nitrous oxide, hydrofluorocarbons, perfluorocarbons, and sulfur hexafluoride—which, for the purposes of accounting, are converted into units known as "carbon dioxide equivalents." Annex 1 nations can meet their targets, in part, by buying and selling emissions "credits" and by investing in "clean development" projects in non–Annex 1 nations, like China and India.

Even as Kyoto was being negotiated, it was clear that the treaty was going to face opposition from many of the same senators who had voted in favor of the original Framework Convention. In July of 1997, Senator Chuck Hagel, Republican of Nebraska, and Senator Robert Byrd, Democrat of West Virginia, introduced a "sense of the Senate" resolution publicly warning the Clinton administration against the direction that the talks were taking. The so-called Byrd–Hagel Resolution stated that the United States should reject any agreement that committed it to reducing emissions unless concomitant obligations were imposed on developing countries as well. The Senate approved the resolution by a vote of 95–0, an outcome that reflected lobbying by both business and labor. (The Global Climate Coalition, a group that was sponsored by, among others, Chevron, Chrysler, Exxon, Ford, General Motors, Mobil, Shell, and Texaco, spent some $13 million on an anti–Kyoto Protocol advertising campaign.)

From a certain perspective, the logic behind the Byrd–Hagel Resolution is unimpeachable. Emissions controls cost money, and this cost has to be borne by somebody. If the United States were to agree to limit its greenhouse gases while economic competitors like China and India did not, then American companies would be put at a disadvantage. "A treaty that requires binding commitments for reduction of emissions of greenhouse gases for the industrial countries but not developing countries will create a very damaging situation for the American economy" is how Richard Trumka, the secretary-treasurer of the AFL–CIO, put it when he traveled to Kyoto to lobby against the protocol. It is also true that an agreement that limits carbon emissions in some countries and not in others could result in a migration, rather than an actual reduction, of CO_2 emissions. (Such a possibility is known in climate parlance as "leakage.")

From another perspective, however, the logic of Byrd-Hagel is deeply, even obscenely, self-serving. Suppose for a moment that the total anthropogenic CO_2 that can be emitted into the atmosphere were a big ice-cream cake. If the aim is to keep global concentrations below five hundred parts per million, then roughly half that cake has already been consumed, and, of that half, the lion's share has been polished off by the industrialized world. To insist now that all countries cut their emissions simultaneously amounts to advocating that industrialized nations be allocated most of the remaining slices, on the ground that they've already gobbled up so much. In a year, the average American produces the same greenhouse-gas emissions as four and a half Mexicans, or eighteen Indians, or ninety-nine Bangladeshis. If both the United States and India were to reduce their emissions proportionately, then the average Bostonian could continue indefinitely producing eighteen times as much greenhouse gases as the average Bangalorean. But why should anyone have the right to emit more than anyone else? At a climate meeting in New Delhi a few years ago, Atal Bihari Vajpayee, then the Indian prime minister, told world leaders, "Our per capita greenhouse gas emissions are only a fraction of the world average and an order of magnitude below that of many developed countries. We do not believe that the ethos of democracy can support any norm other than equal per capita rights to global environmental resources."

Outside the United States, the decision to exempt developing nations from Kyoto's mandates was generally regarded as an adequate—if imperfect—solution to an otherwise intractable problem. The arrangement was basic to the Framework Convention, and it mimicked a structure that had already been employed—successfully—to deal with another potential global crisis: the depletion of atmospheric ozone. The Montreal Protocol, adopted in 1987, called for a phase-out of ozone-depleting chemicals, but gave developing nations what amounted to a ten-year grace period. Pieter van Geel, the Dutch environment secretary, described the European outlook to me as follows: "We cannot say, 'Well, we have our wealth, based on the use of fossil fuels for the last three hundred years, and, now that your countries are growing, you may not grow at this rate, because we have a climate-change problem.' We should show moral leadership by giving the example. That's the only way we can ask something of these other countries."

For its part, the Clinton administration supported the Kyoto Protocol in theory, but not really in practice. In November 1998, the United States's ambassador to the U.N. signed the treaty on behalf of the administration. But the president never submitted it to the Senate, where clearly it wouldn't have won the two-thirds vote needed for ratification. On Earth Day 2000, Clinton delivered more or less the same speech he had given seven years earlier: "The greatest environmental challenge of the new century is global warming. The scientists tell us the 1990s were the hottest decade of the entire millennium. If we fail to reduce the emission of greenhouse gases, deadly heat waves and droughts will become more frequent, coastal areas will flood, and economies will be disrupted. That is going to happen, unless we act." By the time he left office, CO_2 emissions from the United States were 15 percent higher than they had been in 1990.

No politician in America—perhaps no major politician in the world—is more closely associated with the subject of global warming than Al Gore. In 1992, while still in the Senate, Gore published *Earth in the Balance*, in which he argued that protecting the global environment should be the "central organizing principle" of society; five years later, as vice president, he flew to Japan to salvage Kyoto when negotiations seemed on the verge of breaking down. Nonetheless, global warming never really became a factor in the 2000 election. During the campaign, George W. Bush repeatedly asserted that he, too, was deeply concerned about climate change, calling it "an issue that we need to take very seriously." He promised that, if elected, he would impose federal regulations limiting CO_2 emissions.

Soon after his inauguration, Bush sent the new head of the Environmental Protection Agency, Christine Todd Whitman, to a meeting of environmental ministers from the world's leading industrialized nations, where she elaborated on what she apparently believed to be his position. Whitman assured her colleagues that the president considered global warming to be "one of the greatest environmental challenges that we face" and that he wanted to "take steps to move forward." Ten days after her presentation, Bush announced that not only was he withdrawing the United States from the ongoing negotiations over Kyoto—the protocol had left several complex issues of implementation to be resolved later—but also he had changed his mind about federal curbs on carbon dioxide. Explaining this reversal, Bush asserted that he no longer thought CO_2 limits were justified, owing to the "state of scientific knowledge of the causes of, and solutions to, global climate change," which he labeled "incomplete." (Former Treasury Secretary Paul O'Neill, who backed the president's original position, has speculated publicly that the reversal was engineered by Vice President Dick Cheney.)

For nearly a year, the Bush administration operated essentially without any position on climate change. Then, the president announced that the United States would be pursuing a whole new approach. Instead of focusing on greenhouse gas emissions, the country would focus on something called "greenhouse gas intensity." Bush declared this new approach preferable because it recognized "that a nation that grows its economy is a nation that can afford investments and new technology."

Greenhouse gas intensity is not a quantity that can be measured directly. Rather, it is a ratio that relates emissions to economic output. Say, for example, that one year a business produces a hundred pounds of carbon and a hundred dollars' worth of goods. Its greenhouse gas intensity in that case would be one pound per dollar. If the next year that company produces the same amount of carbon but an extra dollar's worth of goods, its intensity will have fallen by one percent. Even if it doubles its total emissions of carbon, a company—or a country—can still claim a reduced intensity provided that it more than doubles its output of goods. (Typically, a country's greenhouse gas intensity is measured in terms of tons of carbon per million dollars' worth of gross domestic product.)

To focus on greenhouse gas intensity is to give a peculiarly sunny account of the U.S. situation. Between 1990 and 2000, U.S. greenhouse gas intensity fell by some 17 percent, owing to several factors, including the shift toward a more service-based economy. Meanwhile, total emissions rose by some 12 percent. (In terms of greenhouse gas intensity, the United States actually performs better than many third world nations, because even though we consume a lot more energy, we also have a much larger GDP.) In February 2002, President Bush set the goal of reducing the country's greenhouse gas intensity by 18 percent over the following ten years. During that same decade, his administration expects the American economy to grow by 3 percent annually. If both expectations are met, overall emissions of greenhouse gases will rise by about 12 percent.

The administration's plan, which relies almost entirely

on voluntary measures, has been characterized by critics as nothing more than a subterfuge—"a total charade" is how Philip Clapp, president of the Washington-based National Environmental Trust, once put it. And certainly, if the goal is to prevent "dangerous anthropogenic interference," then greenhouse gas intensity is the wrong measure to use. (Essentially, the president's approach amounts to following the path of "business as usual.") The administration's response to such criticism has generally been to attack its premise. "Science tells us that we cannot say with any certainty what constitutes a dangerous level of warming and therefore what level must be avoided," Paula Dobriansky has stated. When I asked her how, in that case, the United States could support the aim of averting DAI, she answered by saying—twice—"We predicate our policies on sound science."

Right around the time I went to visit Dobriansky, the chairman of the Senate Committee on Environment and Public Works, James Inhofe, gave a speech on the Senate floor, which he titled "An Update on the Science of Climate Change." In the speech, Inhofe, an Oklahoma Republican, announced that "new evidence" had come to light that "makes a mockery" of the notion that human-induced warming is occurring. The senator, who has called global warming "the greatest hoax ever perpetrated on the American people," went on to argue that this important new evidence was being suppressed by "alarmists" who view anthropogenic warming as "an article of religious faith." One of the authorities that Inhofe repeatedly cited in support of his claims was the fiction writer Michael Crichton.

It was an American scientist—Charles David Keeling—who, in the 1950s, developed the technology to measure CO_2 levels precisely, and it was American researchers who, working on Mauna Loa, first showed that these levels were steadily rising. In the half century since then, the United States has contributed more than any other nation to the advancement of climate science, both theoretically, through the work of climate modelers at places like GISS and NOAA's Geophysical Fluid Dynamics Laboratory, and experimentally, through field studies conducted in the Arctic, the Antarctic, and every continent in between.

At the same time, the United States is also the world's chief purveyor of the work of so-called global-warming skeptics. The ideas of these skeptics are published in books with titles like *The Satanic Gases* and *Global Warming and Other Eco-Myths* and then circulated on the Web by groups like Tech Central Station, which is sponsored by, among others, ExxonMobil and General Motors. While some skeptics' organizations argue that global warming isn't real, or at least hasn't been proved—"Predicting *weather* conditions a day or two in advance is hard enough, so just imagine how hard it is to forecast what our *climate* will be," Americans for Balanced Energy Choices, a lobbying organization funded by mining and power companies, declares on its Web site—others maintain that rising CO_2 levels are actually cause for celebration.

"Carbon dioxide emissions from fossil fuel combustion are beneficial to life on earth," the Greening Earth Society, an organization created by the Western Fuels Association, a utility group, states. Atmospheric levels of 750 parts per million—nearly triple preindustrial levels—are nothing to worry about, the society maintains, because plants like lots of CO_2, which they need for photosynthesis. (Research on this topic, the group's Web site acknowledges, has been "frequently denigrated," but "it's exciting stuff" and provides an "antidote to gloom-and-doom about potential changes in earth's climate.")

In legitimate scientific circles, it is virtually impossible to find evidence of disagreement over the fundamentals of global warming. Naomi Oreskes, a professor of history and science studies at the University of California at San Diego, recently tried to quantify the level of consensus. She conducted a study of more than nine hundred articles on climate change published in refereed journals between 1993 and 2003 and subsequently made available on a leading research database. Of these, she found that 75 percent endorsed the view that anthropogenic emissions were responsible for at least some of the observed warming of the past fifty years. The remaining 25 percent, which dealt with questions of methodology or climate history, took no position on current conditions. Not a single article disputed the premise that anthropogenic warming is under way.

Still, pronouncements by groups like the Greening Earth Society and politicians like Senator Inhofe shape

THE DAY AFTER KYOTO

the public discourse on climate change. And this clearly is their point. A few years ago, pollster Frank Luntz prepared a strategy memo for Republican members of Congress, coaching them on how to deal with a variety of environmental issues. (Luntz, who first made a name for himself by helping to craft Newt Gingrich's "Contract with America," has been described as "a political consultant viewed by Republicans as King Arthur viewed Merlin.") Under the heading "Winning the Global Warming Debate," Luntz wrote, "The scientific debate is closing (against us) but not yet closed. There is still a window of opportunity to challenge the science." He warned, "Voters believe that there is *no consensus* about global warming in the scientific community. Should the public come to believe that the scientific issues are settled, their views about global warming will change accordingly." Luntz also advised, "The most important principle in any discussion of global warming is your commitment to sound science."

It is in this context, and really only in this context, that the Bush administration's claims about the science of global warming make any sense. Administration officials are quick to point to the scientific uncertainties that remain about global warming, of which there are many. But where there is broad agreement, they are reluctant to acknowledge it.

"When we make decisions, we want to make sure we do so on sound science," the president said, announcing his new approach to global warming in February 2002. Just a few months later, the Environmental Protection Agency delivered a two hundred and sixty-three page report to the U.N. that stated, "Continuing growth in greenhouse gas emissions is likely to lead to annual average warming over the United States that could be as much as several degrees Celsius (roughly 3 to 9 degrees Fahrenheit) during the 21st century." The president dismissed the report—the product of years of work by federal researchers—as something "put out by the bureaucracy." The following spring, the EPA made another effort to give an objective summary of climate science, in a report on the state of the environment. The White House interfered so insistently in the writing of the global warming section—at one point, it tried to insert excerpts from a study partly financed by the American Petroleum Institute—that, in an internal memo, agency staff members complained that the section "no longer accurately represents scientific consensus." (When the EPA finally published the report, the climate-science section was missing entirely.) In June 2005, the *New York Times* revealed that a White House official named Philip Cooney had repeatedly edited government reports on climate change in order to make their findings seem less alarming. In one instance, Cooney received a report stating: "Many scientific observations point to the conclusion that the Earth is undergoing a period of relatively rapid change." He revised this statement to read: "Many scientific observations <u>indicate</u> that the Earth <u>may be</u> undergoing a period of relatively rapid change." Shortly after his editing efforts were disclosed, Cooney resigned from his White House post and took a job with Exxon-Mobil.

On the day after the Kyoto Protocol took effect, the United Nations hosted a conference titled, appositely, "One Day After Kyoto." The conference, whose subtitle was "Next Steps on Climate," was held in a large room with banks of curved desks, each equipped with a little plastic earpiece. The speakers included scientists, insurance-industry executives, and diplomats from all over the world, among them the U.N. ambassador from the tiny Pacific island nation of Tuvalu, who described how his country was in danger of simply disappearing. Britain's permanent representative to the U.N., Sir Emyr Jones Parry, began his remarks to the crowd of two hundred or so by stating, "We can't go on as we are."

When the United States withdrew from negotiations over Kyoto, in 2001, the entire effort nearly collapsed. All on its own, America accounts for 34 percent of Annex 1 emissions. According to Kyoto's elaborate ratification mechanism, in order to take effect the protocol had to be approved by countries responsible for at least 55 percent of those emissions. European leaders spent more than three years working behind the scenes, trying to line up support from the remainder of the industrialized world. The crucial threshold was finally crossed in October 2004, when the Russian Duma voted in favor of ratification. The Duma's vote was all but explicitly understood to be a condition of European backing for Russia's bid to join the World

Trade Organization. (RUSSIA FORCED TO RATIFY KYOTO PROTOCOL TO BECOME WTO MEMBER, read the headline in *Pravda*.)

As speaker after speaker at the U.N. conference noted, Kyoto is an important first step, but only a first step. The protocol expires in 2012 and the cuts it mandates don't come close to stabilizing worldwide emissions. Even if every country—including the United States—were to fulfill its obligations under Kyoto, CO_2 concentrations in the atmosphere would still be headed to five hundred parts per million, and beyond. Without substantive commitments from countries like China and India, there is no realistic way to avoid DAI. But why should China and India accept the costs of controlling emissions when America has refused to do so? In this way, the United States, having failed to defeat Kyoto, may be in the process of doing something even more damaging: ruining the chances of reaching a post-Kyoto agreement. "The blunt reality is that, unless America comes back into some form of international consensus, it is very hard to make progress" is how Britain's prime minister, Tony Blair, recently put it.

Astonishingly, standing in the way of this progress seems to be Bush's goal. Dobriansky explained the administration's position to me as follows: While the rest of the industrialized world is pursuing one strategy (emissions limits), the United States is pursuing another (no emissions limits), and it is still too early to say which approach will work best. "It is essential to really implement these programs and approaches now and to take stock of their effectiveness," she said, adding, "we think it is premature to talk about future arrangements." At a round of international climate talks held in Buenos Aires in December 2004, many delegations were pressing for a preliminary round of meetings so that work could start on mapping out a successor to the Kyoto Protocol. The U.S. delegation opposed these efforts so adamantly that finally the Americans were asked to describe, in writing, what sort of meeting they would find acceptable. They issued half a page of conditions, one of which was that the session "shall be a one-time event held during a single day." Another condition was, paradoxically, that, if they were going to discuss the future, the future would have to be barred from discussion; presentations, they wrote, should be limited to "an information exchange" on "existing national policies." Annie Petsonk, a lawyer with the advocacy group Environmental Defense, who previously worked for the administration of George Bush Senior, attended the talks in Buenos Aires. She recalled the effect that the memo had on the members of the other delegations: "They were ashen."

European leaders have made no secret of their dismay at the administration's stance. "It's absolutely obvious that global warming has started," France's president, Jacques Chirac, said after attending the 2004 summit of leaders of the world's major industrial powers—the Group of 8. "And so we have to act responsibly, and, if we do nothing, we would bear a heavy responsibility. I had the chance to talk to the United States president about this. To tell you that I convinced him would be a total exaggeration, as you can imagine." Tony Blair, who held the presidency of the G8 in 2005, spent the months leading up to that year's summit trying to convince Bush that, in his words, "the time to act is now." It's plain, Blair said in an address devoted to climate change, that "the emission of green-house gases . . . is causing global warming at a rate that began as significant, has become alarming, and is simply unsustainable in the long-term. And by 'long-term' I do not mean centuries ahead. I mean within the lifetime of my children certainly; and possibly within my own. And by 'unsustainable,' I do not mean a phenomenon causing problems of adjustment. I mean a challenge so far-reaching in its impact and irreversible in its destructive power, that it alters radically human existence." Just a few weeks before the 2005 summit, which was held in Gleneagles, Scotland, the national science academies of all the G8 nations, including the United States, along with the science academies of China, India, and Brazil, issued a remarkable joint statement calling on world leaders to "acknowledge that the threat of climate change is clear and increasing."

All of this, however, had no apparent impact on the president. In the lead-up to the summit, the head of the White House Council on Environmental Quality, James Connaughton, attended a meeting in London where he announced that he still wasn't convinced that anthropogenic warming was a problem. "We are still working on the issue of causation, the extent to which humans are a factor," he said. According to the *Washington Post*, administration officials insisted on weakening a proposal for joint action prepared for the summit, demanding, for example, the deletion of a passage citing "increasingly compelling evidence of climate change, including rising ocean and atmospheric temperatures, retreating ice sheets and glaciers, rising sea levels, and changes to ecosystems." The final communiqué from the summit, which was overshadowed by the London subway bombings, largely reflected the administration's position; it labeled global warming a "serious and long-term challenge" but also cited "uncertainties" in "our understanding of climate science" and called vaguely on G8 members to "promote innovation" and "accelerate deployment of cleaner technologies."

Senator John McCain, Republican of Arizona, is the primary sponsor of a bill that would, in effect, make good on George Bush's unfulfilled 2000 campaign promise to regulate carbon emissions. The Climate Stewardship Act calls for a reduction of greenhouse gas emissions in the United States to 2000 levels by 2010, and to 1990 levels by 2016. McCain has managed to get the Climate Stewardship Act onto the Senate floor for a vote twice, both times over strong White House opposition. In October 2003, the measure was defeated by a vote of fifty-five to forty-three; in June 2005, it went down sixty to thirty-eight. When I asked McCain to characterize Bush's position on global warming, he responded, "MIA."

"This is clearly an issue that we will win on over time because of the evidence," he went on. "The overwhelming impacts of climate change are becoming more and more visible every day. The problem is: will it be too late? We are a country that emits nearly 25 percent of the world's greenhouse gases. How much damage will have been done before we act?"

As of this writing, U.S. emissions are nearly 20 percent higher than they were in 1990.

Chapter 9

BURLINGTON, VERMONT

BURLINGTON, VERMONT, on the eastern shore of Lake Champlain, is by almost any measure a small city; still, it is the largest in Vermont. Several years ago, its voters decided that instead of authorizing the local utility company to buy more power, they would use less of it. Since then, the city has probably done as much as any municipality in the country to try to reduce its greenhouse gas emissions. The Burlington Electric Department may be the only utility in the United States whose vehicle fleet includes mountain bikes.

Peter Clavelle has been Burlington's mayor since 1989, with a two-year hiatus, which he likes to refer to as a "voter-inspired sabbatical." He is short and bald, with a salt-and-pepper mustache and mournful blue eyes. During his "sabbatical," Clavelle went to live with his family on the island of Grenada.

"Living on an island, you really get in touch with practices that are sustainable and practices that are unsustainable," he told me. It was a sticky July day, and we were driving around town in Clavelle's hybrid Honda Civic, looking at energy-saving projects. He paused to point out a city bus equipped with a bicycle rack on the front grille.

"The issues around climate protection are about sustainability," he went on. "They're about future generations. They're also about this conviction that local action does make a difference. Many of us are very frustrated with the lack of vision and action by the federal government, but there's a choice to be made. You either can bemoan federal policies or you can take control of your own destiny."

Burlington's energy-saving campaign, launched in 2002, is known as the "10 percent challenge." ("Put the chill on global warming" is its slogan.) As the name of the campaign suggests, the city's aim is to reduce greenhouse gas emissions by 10 percent, though from what baseline is somewhat vaguely defined. To further this goal, Burlington has tried just about everything, from providing free energy consultations to businesses to designing "energy efficiency calendars" for kids. Tray liners printed up for the local McDonald's feature a well-meaning but creepy-looking dinosaur named Climo Dino. "While our climate was changed by a giant asteroid, you humans are changing your own climate by emitting six billion tons of CO_2 into the atmosphere each year," Climo Dino observes.

The first stop on Clavelle's tour was an outpost of the county dump where, instead of collecting rubbish, the city sells it. Burlington encourages contractors to engage not in demolition but in "deconstruction," a practice that saves energy both by reducing the city's waste stream and cutting down on the need for new materials. Dozens of "deconstructed" sinks and doors and vanities were arrayed, showroom style, in what once had been a garage. A next-to-new staircase was leaning against the wall, waiting for a buyer needing steps of precisely the same dimensions. In the parking lot, some kids were building a garden shed out of old plywood. Clavelle told me that he had gotten the idea for ReCycle North from a similar program in Minneapolis. "It's management by plagiarism," he announced cheerfully.

Our next stop was the headquarters of the Burlington Electric Department, or BED for short. Behind the building, I could see a single wind turbine, which was turning briskly in the breeze. The turbine symbolizes the city's effort, while at the same time generating enough power for thirty homes. All in all, Burlington Electric gets nearly half of its energy from renewable sources, including a fifty-megawatt power plant in town that runs off wood chips. As we headed inside the BED headquarters, we passed a display of compact fluorescent light bulbs, which the company leases to interested customers at a cost of twenty cents a month. An electric department official named Chris Burns came out to greet us. He explained that a family that was keeping a hundred-watt incandescent bulb burning out on the porch all night could cut its electricity bill by up to 10 percent by simply replacing that bulb with a compact fluorescent. He said that several businesses in Burlington had cut their energy usage by significantly more than that just by taking such basic steps as adjusting the thermostat. The Burlington Electric Department has estimated that the energy-saving projects that the city has undertaken will, over the course of their useful life, prevent the release of nearly 175,000 tons of carbon. "We consider every building a power plant," Burns told me.

A little later in the day, Clavelle took me to visit the City Market, a grocery store built on municipal land that had previously been a hazardous waste site. The city supports the market so that Burlington residents won't have to drive to the suburbs to go food shopping. The market, in turn, is heavily stocked with local produce. "We estimated that a typical tomato traveled twenty-five hundred miles to reach our kitchen table," Clavelle said. "And we could produce that tomato right here." Finally, we headed over to a section of town known as the Intervale. A flood plain along the Winooski River, the Intervale once was a farming district, then it was a wasteland, and now it is home to an assortment of community gardens and cooperatives with names like the Lucky Ladies Egg Farm and the Stray Cat Farm. By the time we arrived at the Intervale, the weather had changed for the worse. In the pouring rain, we stopped at an old brick farmhouse. Summer squash of various shapes and sizes were displayed out front. Next door was a composting facility that collects vegetable waste from local restaurants and turns it back into soil.

"It's a closed loop," Clavelle told me.

One consequence, presumably unintended, of America's failure to ratify the Kyoto Protocol has been the emergence of a not–quite-grassroots movement. In February 2005, Greg Nickels, the mayor of Seattle, began to circulate a set of principles that he called the "U.S. Mayors Climate Protection Agreement." Within four months, more than a hundred and seventy mayors, representing some thirty-six million people, had signed on, including Mayor Michael Bloomberg of New York; Mayor John Hickenlooper of Denver; and Mayor Manuel Diaz of Miami. Signatories agreed to "strive to meet or beat the Kyoto Protocol targets in their own communities." At around the same time, officials from New York, New Jersey, Delaware, Connecticut, Massachusetts, Vermont, New Hampshire, Rhode Island, and Maine announced that they had reached a tentative agreement to freeze power plant emissions from their states at current levels and then begin to cut them. Even Governor Arnold Schwarzenegger, the Hummer collector, joined in; an executive order he signed in June 2005 called on California to reduce its greenhouse gas emissions to 2000 levels by 2010 and to 1990 levels by 2020. "I say the debate is over," Schwarzenegger declared right before signing the order.

Burlington's experience demonstrates how much can, indeed, be accomplished through local action. In the sixteen years since Clavelle became mayor, electricity usage in the state of Vermont has risen by nearly 15 percent. In Burlington, by contrast, it has dropped by one percent. The savings were achieved entirely through voluntary measures, by homeowners and businesses who, presumably, came to see controlling their utility bills as in their own self-interest.

But Burlington's experience also makes the limits of local action obvious. The biggest reductions were achieved early on, when the city approved a bond issue to fund energy conservation projects. As the most inefficient homes and businesses in the city were upgraded, gains became harder and harder to come by. Since the 10 percent challenge was initiated, in 2002, electricity demand in the city has actually started to creep back up again and is now slightly higher than it was at the campaign's launch. Meanwhile, whatever savings have been made in electricity usage have been offset by increased CO_2 emissions from other sources, mostly cars and trucks. As we were heading back to City Hall, I asked Clavelle what more could be done.

"It would be so much easier if we could say, 'Well, if we approved this one project or this action, the problem would be solved,'" he told me. "But there's no silver bullet. There's no one thing we can do. There's no *ten* things we can do. There's hundreds and hundreds of things that we need to do.

"I'm frustrated," he said. "But you need to remain hopeful."

The headquarters of the Natural Resources Defense Council are situated on West Twentieth Street in Manhattan. The offices, which occupy the top three floors of a twelve-story art deco building, were designed in 1989 as a prototype for energy-efficient urban life, with "occupancy sensors" that shut off the lights automatically when no one's around and special polymer-coated windows that help keep out heat. A large skylight above the staircase is supposed to provide natural light to the reception area, though after fifteen years the glass has been coated with a fine layer of New York grime.

David Hawkins runs NRDC's climate program. He is tall and thin, with dark, wavy hair and a gentle manner. Hawkins joined the environmental group thirty-five years ago, fresh out of law school, and has worked there ever since, with one break, in the late 1970s, when he served as head of the EPA's air quality division. These days, he spends a lot of his time in China, meeting with officials at places like the National Development and Reform Commission and the Shanxi Institute of Coal Chemistry.

Over the next fifteen years, the size of China's economy is expected to more than double. This projected growth, most of which will be fueled by coal, overwhelms not just all conservation projects that are currently being undertaken in the United States, but also any that could be reasonably imagined. Hawkins gave me a copy of a presentation he had prepared on future power plant construction. In it was a graph detailing China's plans: by 2010, the country is expected to build 150 new one thousand-megawatt coal plants (or their generating equivalent); by 2020, it is expected to construct another 168 new plants. If every single town and city in the United States were to match the efforts that Burlington has made, the aggregate savings would amount—very roughly—to 1.3 billion tons of carbon over the next several decades. Meanwhile, the lifetime emissions just from the new coal plants China is expected to build would amount to some 25 billion tons of carbon. To put this somewhat differently, China's new plants would burn through all of Burlington's savings—past, present, and future—in less than two and a half hours.

Despair might seem the logical response to such figures. In this way, the hazard of looking objectively at global warming can be almost as great as refusing to see the problem at all. Hawkins, though, is an optimist—perhaps by professional necessity. "If you're looking at global warming, you look at what the emissions are from the large industrial and industrializing countries," he told me. "And it doesn't take very long to conclude that you can't solve this problem unless you deal with the United States and China, and if you deal with the United States and China, you can solve this problem."

"China is in the takeoff stage," he went on. "So there's an opportunity to build things there using modern technology rather than to build them using pickup technology. And that's the challenge for us: to do things that convince the Chinese that that's the better strategy for them."

Right now, he pointed out, China is industrializing according to a model set in the United States forty or fifty years ago: its factories rely on obsolete and highly inefficient motors; its electricity transmission system is antiquated; and although it is the world's primary manufacturer of compact fluorescent bulbs, it barely uses any. (Per unit of gross domestic product, China consumes two and a half times as much energy as the United States and nearly nine times as much as Japan.) Were China to bring its factories up to date and fill even a modest amount of its projected energy demand from renewable sources, it is estimated that the number of new coal-fired plants it would need to build could be cut by nearly a third.

At this point, China is building only conventional coal-fired plants. For technical reasons, "carbon capture and storage," or CCS, isn't feasible with this type of plant. But if China were to shift to a method known as coal gasification, then—potentially at least—the CO_2 emissions from the new plants could be captured and sequestered. In that case, their carbon emissions would be substantially lower—possibly zero. It is estimated that together, coal gasification technology and carbon capture and storage would add 40 percent to the costs of a new plant. (This is an imprecise figure, since CCS has never actually been tried at a commercial power plant.) Hawkins has calculated that even assuming such a high differential cost, the added expense of carbon capture and storage for all the new coal plants expected to be built in all of the world's developing nations could be paid for through a one percent tax on the electricity bills of consumers in developed nations. "So it is affordable," he told me.

China's growth is often cited as a justification for U.S. inaction. What's the point of going to a lot of trouble if, in the end, your efforts won't make a difference? Hawkins maintains that this argument gets things completely backward. What America does, China in the long run will do too. "This isn't theory," he said. "We saw it with automobile pollution controls. We adopted those in the seventies and those modern pollution controls are being required around the world today. Sulfur dioxide scrubbers on power plants—we applied them; China is now applying them. There's a very practical reason why this works, and that is if a country like the United States embraces a cleanup technology, then the market starts to drive the price down, and other countries start to see that it is doable." Although no new coal-fired power plants have been built in the United States in recent years, many analysts expect this to change in the coming decade. Hawkins argues that American utilities should be prohibited from constructing any new plants without CCS capability.

"If we can get policies adopted that prevent the U.S. from building new coal plants that don't capture their emissions and create incentives for the Chinese to build new coal plants that will capture their emissions, then it doesn't matter if there's an international treaty or not," he said. "If we get the facts on the ground right, we've bought time."

Chapter 10

MAN IN THE ANTHROPOCENE

A FEW YEARS AGO, in an essay in *Nature*, the Nobel Prize–winning Dutch chemist Paul Crutzen coined a term. No longer, he wrote, should we think of ourselves as living in the Holocene. Instead, an epoch unlike any of those which preceded it had begun. This new age was defined by one creature—man—who had become so dominant that he was capable of altering the planet on a geological scale. Crutzen dubbed this age the "Anthropocene."

Crutzen's was not the first such neologism. Already in the 1870s, the Italian geologist Antonio Stoppani argued that human influence was ushering in a new age, which he called the "anthropozoic era." A few decades later, the Russian geochemist Vladimir Ivanovich Vernadsky proposed that the earth was entering a new stage—the "noosphere"—dominated by human thought. But while these earlier terms had had a positive slant—"I look forward with great optimism . . . We live in a transition to the noosphere," Vernadsky wrote—the connotations of the Anthropocene were distinctly cautionary. Humans had become the driving agents on the planet, yet it wasn't at all clear they knew where they were going.

Crutzen won the Nobel for his work on the chemistry of ozone depletion, a phenomenon that offers many parallels, both scientific and social, to global warming. The most prevalent ozone-destroying chemicals—chlorofluorocarbons—are odorless, colorless, nonreactive, and, much like CO_2, apparently benign. (To demonstrate their safety, their inventor once inhaled some CFCs and then used the vapors to blow out a set of birthday candles.) Starting in the 1930s, the "wonder gas" was employed as a refrigerant and in the 1940s as an ingredient in Styrofoam. The first indication that chlorofluorocarbons were anything to worry about didn't come until the 1970s, when research chemists began to consider—purely as an academic exercise—what would happen to CFCs in the upper atmosphere. They determined that although the chemicals were stable near the earth's surface, in the stratosphere they wouldn't be. Once CFCs started to break down, the result would be free chlorine, which, they hypothesized, would work as a catalyst to convert ozone, O_3, into ordinary oxygen, O_2. Because stratospheric ozone shields the earth from ultraviolet radiation, the researchers warned that continued use of CFCs could have disastrous consequences. F. Sherwood Rowland, who shared the Nobel Prize with Crutzen, came home one night and told his wife, "The work is going well, but it looks like it might be the end of the world."

The damaging effects of CFCs were confirmed—rather

more dramatically than researchers had anticipated—in the 1980s by the discovery that a "hole" had opened up in the ozone layer over Antarctica. (Confirmation might have come earlier had NASA computers not been programmed to reject as erroneous any data on ozone levels that seemed too low.) Even as evidence of chlorofluorocarbons' effects accumulated, American chemical manufacturers, who supplied more than a third of the world's CFCs, continued to resist regulation, arguing on the one hand that more study of the problem was needed and on the other that only unified global action could address it. At one point, President Reagan's interior secretary, Donald Hodel, suggested that if CFCs were indeed destroying the ozone layer, then people should simply wear sunglasses and buy hats. "People who don't stand out in the sun—it doesn't affect them," he asserted. Finally, in 1987, the Montreal Protocol was agreed to, and the process of phasing out CFCs began. (Chlorofluorocarbons, it should be noted, are also a greenhouse gas.) It is expected that sometime in the next several years, ozone levels will bottom out and then begin to creep back up again. Depending on how you look at things, this resolution represents either a triumph of science, or just the reverse. As Crutzen himself has observed, if chlorine had turned out to behave just slightly differently in the upper atmosphere, or if its chemical cousin bromine had been used in its stead, then by the time anyone had thought to look into the state of the ozone layer, the "ozone hole" would have stretched from one pole to the other.

"More by luck than by wisdom this catastrophic situation did not develop," he has written.

In the case of global warming, a much longer interval separates theory and observation. According to Crutzen, the Anthropocene began all the way back in the 1780s, the decade in which James Watt perfected his steam engine. Arrhenius undertook his pen and paper calculations in the 1890s. The retreat of the Arctic sea ice, the warming of the oceans, the rapid shrinking of the glaciers, the redistribution of species, the thawing of the permafrost—these are all new phenomena. It is only in the last five or ten years that global warming has finally emerged from the background "noise" of climate variability. And even so, the changes that can be seen lag behind the changes that have been set in motion. The warming that has been observed so far is probably only about half the amount required to bring the planet back into energy balance. This means that even if carbon dioxide were to remain stable at today's levels, temperatures would still continue to rise, glaciers to melt, and weather patterns to change for decades to come.

But CO_2 levels are *not* going to remain stable. Just to slow the growth, as Socolow and Pacala's "wedge" scheme illustrates, is a hugely ambitious undertaking, one that would require new patterns of consumption, new technologies, and new politics. Whether the threshold for "dangerous anthropogenic interference" is 450 parts per million of CO_2 or 500, or even 550 or 600, the world is rapidly approaching the point at which, for all practical purposes, the crossing of that threshold will become impossible to prevent. To refuse to act, on the grounds that still more study is needed or that meaningful efforts are too costly or that they impose an unfair burden on industrialized nations, is not to put off the consequences, but to rush toward them. The British magazine *New Scientist* recently ran a global warming Q&A, which ended with the question, "How worried should we be?" The answer was another question: "How lucky do you feel?"

Luck and resourcefulness are, of course, essential human qualities. People are always imagining new ways to live, and then figuring out ways to remake the world to suit what they've imagined. This capacity has allowed us, collectively, to overcome any number of threats in the past, some imposed by nature and some by ourselves. It could be argued, taking this long view, that global warming will turn out to be just one more test in a sequence that already stretches from plague and pestilence to the prospect of nuclear annihilation. If, at this moment, the bind that we're in seems insoluble, once we've thought long and hard enough about it we'll find—or perhaps, float—our way clear.

But it's also possible to take an even longer view of the situation. The climate record provided by Greenland ice cores gives a highly resolved history going back more than a hundred thousand years and the Antarctic cores a history stretching back more than four hundred thousand years. What these records show, in addition to a clear correlation between CO_2 levels and global temperatures, is that the last glaciation was a time of frequent and traumatic climate swings. During that period, humans who were, genetically speaking, just like ourselves wandered the globe, producing nothing more permanent than isolated cave paintings and large piles of mastodon bones. Then, ten thousand years ago, the weather changed. As the climate settled down, so did we. People built villages, towns, and, finally, cities, along the way inventing all the basic technologies—agriculture, metallurgy, writing—that future civilizations would rely upon. These developments would not have been possible without human ingenuity, but, until the climate cooperated, ingenuity, it seems, wasn't enough.

Ice core records also show that we are steadily drawing closer to the temperature peaks of the last interglacial, when sea levels were some fifteen feet higher than they are today. Just a few degrees more and the earth will be hotter than it has been at any time since our species evolved. The feedbacks that have been identified in the climate system—the ice-albedo feedback, the water vapor feedback, the feedback between temperatures and carbon storage in the permafrost—take small changes to the system and amplify them into much larger forces. Perhaps the most unpredictable feedback of all is the human one. With six billion people on the planet, the risks are everywhere apparent. A disruption in monsoon patterns, a shift in ocean currents, a major drought—any one of these could easily produce streams of refugees numbering in the millions. As the effects of global warming become more and more difficult to ignore, will we react by finally fashioning a global response? Or will we retreat into ever narrower and more destructive forms of self-interest? It may seem impossible to imagine that a technologically advanced society could choose, in essence, to destroy itself, but that is what we are now in the process of doing.

AFTERWORD

Hurricane Katrina struck New Orleans on the morning of August 29, 2005. The following month, Hurricane Rita made landfall between Sabine Pass, Texas, and Johnson's Bayou, Louisiana; and a month after that, Hurricane Wilma—at one point, the most intense hurricane ever recorded—slammed into the Yucatán Peninsula, just north of Playa del Carmen. Although the Atlantic hurricane season officially ends on November 30, the storms kept coming. Eventually, the National Hurricane Center ran out of names and had to turn to Greek letters. Tropical Storm Zeta formed on December 30 and persisted into the new year. All told there were twenty-seven named storms in 2005, a record. Of these, fifteen grew into full-blown hurricanes, another record. Typically, three or four Category 5 hurricanes form in the North Atlantic over the course of a decade; in 2005, there were three in the course of a single season. Needless to say, this was also a record.

Following Katrina, I made several trips to Louisiana to report on the devastation. During one of them, I drove with some NOAA officials to the tip of Plaquemines Parish, the long, skinny leg of land that twists out into the Gulf of Mexico southeast of New Orleans. Plaquemines is protected by levees along both edges, but during Katrina and again during Rita water came in from all sides. As we headed south, following the Mississippi, we passed rotting piles of fish, citrus groves dying from saltwater exposure, and boats that had been deposited, willy-nilly, on the shoulder of the highway. The further we drove, the more complete the destruction. In Port Sulphur, most houses had been reduced to construction debris. The few buildings still standing had lost their outer walls, so that you could look right through, into what had once been kitchens and living rooms and dens. The trees were festooned with an astonishing assortment of household items: jackets, tires, chairs, bicycles. It became something of a contest to see who could find the most grotesque item in the foliage. Except for an occasional Humvee full of National Guard members and some Spanish-speaking workers, we were the only people in the area. It was eerily still. Finally, near Empire, someone spotted a dead cow draped over a branch like a blanket. Its body was hollowed out and desiccated, its head enormous and bloated.

Hurricanes draw their strength from the warm surface waters of the ocean. This is why they arise only in the tropics and only during the season when sea temperatures are highest. Global warming could be expected to lead to an increase in hurricane intensity, but hurricanes require precise conditions to form—too much wind shear, for example, and they rip apart—so making long-range predictions is complicated. In 2004, researchers at NOAA's Geophysical Fluid Dynamics Laboratory published the results of a study that ran hurricane simulations through nine different climate models. They forecast that over the next few decades there would be an increase in hurricane strength but that it would be modest—indeed, barely observable. Then, in 2005—as as it happened, just weeks before Katrina—a researcher at MIT named Kerry Emanuel published a study of actual storms. The study, which relied on data collected by aircraft, showed that over the last thirty years, the power of hurricanes had already almost doubled. A few weeks later, just before Rita, scientists at Georgia Tech published another study, this one using data collected by satellites. Like Emanuel, the Georgia Tech group found that the models had failed to capture the impact of warming on storm intensity. Between 1975 and 2004, tropical sea-surface temperatures rose by roughly one degree. During that same period, the proportion of hurricanes reaching Category 4 or Category 5 status increased by nearly 100 percent.

In writing about climate change, it's important to acknowledge the many uncertainties that exist. But it's also important to recognize that the uncertainty cuts both ways. "If anything, the history of climate modeling has been one of conservatism and *underestimating* the impacts of climate change," Ken Caldeira, a researcher at the

Carnegie Institution Department of Global Ecology at Stanford University, observed recently. Shortly before he set off for his seventeenth field season at Swiss Camp, Konrad Steffen talked to me about the latest data from Greenland. He said that changes on the ice sheet were occurring an order of magnitude faster than he had been taught to expect.

If the trends are "not sustaining, then we have a problem," he told me. If they are sustaining, "then we have a deep problem."

Since this book first went to press, in the fall of 2005, dozens of important new studies on global warming have appeared. An uncomfortably large proportion of them point to the same conclusion: the world is changing more quickly and more dramatically than had been anticipated. Here are some examples:

● In September 2005, at the end of the summer melt season, satellite measurements showed that the extent of the Arctic ice cap had shrunk to the lowest level ever recorded. The loss was so great that it prompted scientists to revise their forecasts about the ice cap's future. While earlier they had predicted that the Arctic Ocean could be ice-free in summer by 2080, now they predict that it could be ice-free "well before the end of this century." Sea ice begins to grow as the Arctic days become shorter, but in April 2006, the National Snow and Ice Data Center reported that its extent was also the lowest ever recorded at the end of winter.

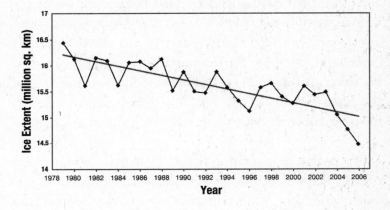

Winter sea ice extent has dropped sharply in recent years.
Credit: National Snow and Ice Data Center, 2006.

• A group of marine biologists, writing in *Nature*, warned that due to the build up of CO_2 in the oceans, key marine organisms could have difficulty surviving. (Carbon dioxide dissolved in seawater forms a weak acid, which interferes with the process of shell formation.) Detrimental conditions, the group warned, "could develop within decades, not centuries as suggested previously."

• NOAA reported that atmospheric CO_2 levels jumped by a near-record 2.53 parts per million in 2005. "The rate of increase is accelerating," Pieter Tans, NOAA's chief carbon dioxide analyst, said.

• A study by researchers at NASA and the University of Kansas concluded that between 1996 and 2005 the loss of ice from Greenland doubled, as glacier flow accelerated.

• Analyzing data from satellites that can detect minute changes in the earth's gravitational field, scientists at the University of Colorado concluded that Antarctica has been losing ice. The scientists put this loss at about thirty-six cubic miles per year. The finding was particularly ominous because climate modelers had expected the overall mass of the Antarctic ice sheet to *increase*, as rising temperatures led to more snowfall over the continent's midsection.

Even as the news about the climate has grown more urgent, the situation in Washington has remained essentially static. The Bush administration continues to claim that the United States is committed to avoiding DAI and it continues to block all efforts to achieve this goal. In December 2005, the eleventh conference of the parties to the U.N. Framework Convention took place in Montreal. When the discussion turned to initiating talks on a post-Kyoto treaty, the chief climate negotiator for the United States, Harlan Watson, literally walked out of the meeting. At a news conference in Montreal, Under Secretary of State Paula Dobriansky was asked how she would answer "those in the United States who are saying you are not doing enough?" She responded—amazingly enough—"We act, we learn, we act again."

The administration similarly continues to suppress information it doesn't care to hear. In early 2006, NASA's James Hansen told reporters that the White House had tried to censor him by insisting that the agency's public affairs staff vet his lectures and papers. (Shortly after this, one of the officials who had tried to limit Hansen's remarks, a twenty-four-year-old political appointee named George Deutsch, was shown to have falsified his résumé; he subsequently resigned.) A few weeks later, it was revealed that the president had turned to Michael Crichton, whose thriller *State of Fear* portrays climate change as fiction invented by environmentalists, for advice on how to deal with the issue. Bush and Crichton reportedly "talked for an hour and were in near-total agreement."

Because U.S. participation is crucial to a post-Kyoto treaty, a great deal clearly hinges on the upcoming presidential election. There are many who argue that even though U.S. policy appears to be stalled, attitudes among voters are changing. I don't entirely share this view, but certainly some signs are encouraging. Over the past year, a number of religious groups and major corporations have begun to press for mandatory emissions limits. Meanwhile, the list of states and municipalities trying to act on their own continues to grow. (To cite an example very close to home, residents of the small western Massachusetts town where I live voted a few months ago to offer tax breaks to families that purchase fuel-efficient cars.) After his movie *An Inconvenient Truth* grossed more per screen on its opening weekend than any other documentary in history, Al Gore pronounced himself "optimistic" that the United States would respond in time. He likened the political system to the climate: "It's nonlinear. It can appear to move at a glacier's pace and then, after crossing a tipping point, it can suddenly move rapidly into a completely new pattern. I've seen that happen."

In Louisiana, I spoke with a lot of people who were living in trailers because their homes and everything in them had been destroyed. I also talked to engineers and hydrologists, levee supervisors and surveyors. What they had to say was eerily familiar. In the years leading up to Katrina, just about everyone who studied New Orleans's situation came to the same conclusion: it was untenable. Owing to its geology, the city was sinking. A wide variety of human activities—from pumping out groundwater to digging channels for oil exploration—were exacerbating the problem. The wetlands that had once surrounded New Orleans, helping to buffer it from storm surges, were degraded, disappearing at the rate of a football field every thirty-eight minutes. Even by the Army Corps of Engineers' own reckoning, the levees built to protect the city were insufficient, and the Corps was known to be systematically understating the risks. In October 2001, *Scientific American* ran a story that described New Orleans as "a disaster waiting to happen," and warned that "only massive reengineering of southeastern Louisiana can save the city." The following year, the *Times-Picayune* published a five-part, fifty-thousand-word series, titled "Washing Away," which carried much the same message. The official publication of the American Society of Civil Engineers asked: "Can any defense ultimately protect a city that is perpetually sinking?"

All of the studies and news stories were there for everyone to read. But the storm of the future lay in the future, while the costs of preparing for it would have had to be borne in the present. It was easier, both psychically and economically, to turn away from the facts. And so life went on as before, and everyone hoped for the best.

Williamstown, MA
July 2006

CHRONOLOGY

1769: James Watt patents his steam engine.
Atmospheric CO_2 levels are ~ 280 parts per million.

1859: John Tyndall builds the world's first ratio spectrophot-ometer and tests the absorptive properties of atmospheric gases.

1895: Svante Arrhenius completes his calculations on varying CO_2 levels.
Atmospheric CO_2 levels are ~ 290 parts per million.

1928: CFCs are invented.

1958: CO_2 measuring equipment is installed at the Mauna Loa Observatory.

1959: CO_2 levels stand at 315 parts per million.

1970: Paul Crutzen warns that human actions may damage ozone layer.

1979: The National Academy of Sciences issues its first major report on global warming: "We may not be given a warning until the CO_2 loading is such than an appreciable climate change is inevitable."
CO_2 levels reach 337 parts per million.

1987: The Montreal Protocol is adopted; phaseout of CFCs begins.

1988: The Intergovernmental Panel on Climate Change is established by the World Meteorological Organization and the United Nations Environment Programme.

1992: President George H. W. Bush signs the U.N. Framework Convention on Climate Change in Rio de Janeiro.
The U.S. Senate approves the Framework Convention by unanimous consent.
CO_2 levels reach 356 parts per million.

1995: The Intergovernmental Panel on Climate Change issues its Second Assessment Report: "The balance of evidence suggests a discernible human influence on global climate."

1997: The Kyoto Protocol is drafted.

1998: Average global temperatures for the year are the warmest on record.

2000: Presidential candidate George W. Bush calls global warming an "issue that we need to take very seriously."
CO_2 levels are measured at 369 parts per million.

2001: The IPCC issues its Third Assessment Report: "Most of the warming observed over the last fifty years is attributable to human activities."
A report by the National Research Council requested by President Bush states, "Greenhouse gases are accumulating in Earth's atmosphere as a result of human activities, causing surface air temperatures and subsurface ocean temperatures to rise. Temperatures are, in fact, rising."
President Bush announces that the United States is withdrawing from the Kyoto Protocol.
Third warmest year on record.

2002: Larsen B ice shelf collapses.
Second warmest year on record, tied with 2003.

2003: Senator James Inhofe, chairman of the Committee on Environment and Public Works, says he has "compelling evidence that catastrophic global warming is a hoax."
The American Geophysical Union issues a consensus

statement asserting: "Natural influences cannot explain the rapid increase in global near-surface temperatures." CO_2 levels reach 375 parts per million.

2004: Kyoto Protocol is ratified by Russia.
Fourth warmest year on record.

2005: Extent of melt on the Greenland ice sheet reaches a record maximum.
Arctic sea ice reaches a record minimum; researchers warn sea could be ice-free in summer "well before the end of this century."
Kyoto Protocol goes into effect.
The National Academies of Sciences of the eight major industrialized nations issue a joint statement: "The scientific understanding of climate change is now sufficiently clear to justify nations taking prompt action."
The Atlantic hurricane season sets a record for the number of Category 5 Storms
Average global temperatures are statistically tied with 1998.

2006: CO_2 levels reach 381 parts per million. Annual rise is a near-record 2.53 parts per million.
Researchers report that since 1996, the loss of ice from Greenland has doubled.

ACKNOWLEDGMENTS

M ANY VERY BUSY people gave generously of their time and their expertise to make this book possible. A number of them have been named in the preceding pages, but a number have not.

I'd like to thank Tony Weyiouanna, Vladimir Romanovsky, Glenn Juday, Larry Hinzman, Terry Chapin, Donald Perovich, Jacqueline Richter-Menge, John Weatherly, Gunter Weller, Deborah Williams, Konrad Steffen, Russell Huff, Nicolas Cullen, Jay Zwally, Oddur Sigurdsson, and Robert Correll for the help they provided on the chapters concerning the Arctic.

Similarly, I am indebted to Chris Thomas, Jane Hill, William Bradshaw, and Christina Holzapfel for their explication of evolutionary biology; to James Hansen, David Rind, Gavin Schmidt, and Drew Shindell for their lessons in climate modeling; and to Harvey Weiss and Peter deMenocal for sharing their work on ancient civilizations. Pieter van Geel, Pier Vellinga, Wim van der Weegen, Chris Zevenbergen, Dick van Gooswilligen, Jos Hermsen, Hendrik Dek, and Eelke Turkstra were extremely gracious to me when I visited the Netherlands. Robert Socolow, Stephen Pacala, Marty Hoffert, David Hawkins, Barbara Finamore, and Jingjing Qian spent many hours with me discussing mitigation strategies, while Senator John McCain, former vice president Al Gore, Annie Petsonk, James Mahoney, and Under Secretary of State Paula Dobriansky helped me to understand the politics of global warming. Mayor Pete Clavelle kindly showed me around Burlington. Michael Oppenheimer, Richard Alley, Daniel Schrag, and Andrew Weaver were always willing—and able—to answer one last question.

This book began as a series of pieces that appeared in the *New Yorker* magazine. I am deeply grateful to David Remnick for urging—indeed compelling—me to write those pieces. I also want to thank Dorothy Wickenden and John Bennet, who offered much valuable advice; Michael Specter, who provided ideas and encouragement along the way; Louisa Thomas, who generously and ably helped with research; Elizabeth Pearson-Griffiths and Maureen Klier, who copyedited the chapters; and Marisa Pagano, who kindly assisted with the illustrations. I am indebted, too, to Greg Villepique and Yelena Gitlin, who worked so hard to make this book happen.

Gillian Blake and Kathy Robbins guided this project to completion. I am grateful to both of them for their insight and support.

Finally, I want to thank my husband, John Kleiner, who helped in more ways than he should have. Without his peculiar optimism, not a word here would have been written.

SELECTED BIBLIOGRAPHY
AND NOTES

Most of the information contained in this book either comes from interviews or is part of the general—and vast—climate science literature. I have also cited or relied on a number of individual reports, articles, and earlier books, some of which are listed below.

Chapter 1: Shishmaref, Alaska

A study commissioned by the U.S. Army Corps of Engineers, "Shishmaref Relocation and Collocation Study: Preliminary Costs of Alternatives," December 2004, provides detailed information on the village's proposed move.

The official title of the Charney Report is "Report of an Ad Hoc Study Group on Carbon Dioxide and Climate: A Scientific Assessment to the National Academy of Sciences" (Washington, D.C.: National Academy of Sciences, 1979).

Global temperature data for the last two thousand years are drawn from Michael E. Mann and Philip D. Jones, "Global Surface Temperatures over the Past Two Millennia," *Geophysical Research Letters*, vol. 30, no. 15 (2003).

Figures on methane releases from the Stordalen mire are taken from Torben R. Christensen et al., "Thawing Sub-Arctic

Permafrost: Effects on Vegetation and Methane Emissions,"
Geophysical Research Letters, vol. 31, no. 4 (2004).
An account of the mission of the *Des Groseilliers* can be found in
D. K. Perovich et al., "Year on Ice Gives Climate Insights,"
Eos (Transactions, American Geophysical Union), vol. 80, no.
481 (1999).
Figures on the thinning of the Arctic sea ice come from D. A.
Rothrock et al., "Thinning of the Arctic Sea-Ice Cover,"
Geophysical Research Letters, vol. 26, no. 23 (1999).
A fuller discussion of the orbital changes and timing of ice ages can
be read in John Imbrie and Katherine Palmer Imbrie, *Ice Ages:
Solving the Mystery*, revised edition (Cambridge, MA: Harvard
University Press, 1986).

Chapter 2: A Warmer Sky

A useful primer on the science of global warming is John
Houghton, *Global Warming: The Complete Briefing*, third edition
(Cambridge: Cambridge University Press, 2004).
The history of the science of global warming is related in Spencer
R. Weart, *The Discovery of Global Warming* (Cambridge, MA:
Harvard University Press, 2003); and Gale E. Christianson,
Greenhouse: The 200-Year Story of Global Warming (New York:
Walker and Company, 1999). Also the Tyndall Centre for
Climate Change Research offers a detailed biography of its
namesake on its Web site, http://www.tyndall.ac.uk.
John Tyndall's dying words, as recalled by his wife, are recounted
in Mark Bowen, *Thin Ice* (New York: Henry Holt, 2005).
Svante Arrhenius's predictions for better living through CO_2 are
from *Worlds in the Making: The Evolution of the Universe* (New
York: Harper, 1908).
Charles David Keeling wrote about "having fun" trying to
measure CO_2 in his essay "Rewards and Penalties of Monitoring the Earth," *Annual Review of Energy and the Environment*,
vol. 23 (1998).

Chapter 3: Under the Glacier

An excellent account of what's been learned from the Greenland ice is Richard B. Alley, *The Two-Mile Time Machine: Ice Cores, Abrupt Climate Change, and Our Future* (Princeton: Princeton University Press, 2000).

Figures on the acceleration of the Greenland ice sheet are from H. Jay Zwally et al., "Surface Melt–Induced Acceleration of Greenland Ice-Sheet Flow," *Science*, vol. 297 (2002).

Figures on the acceleration of the Jakobshavn Isbrae can be found in W. Abdalati et al., "Large Fluctuations in Speed on Greenland's Jakobshavn Isbrae Glacier," *Nature*, vol. 432 (2004).

James E. Hansen wrote about the future of the Greenland ice sheet in his essay "A Slippery Slope: How Much Global Warming Constitutes 'Dangerous Anthropogenic Interference'?" *Climatic Change*, vol. 68 (2005).

A thorough discussion of abrupt climate change is *Abrupt Climate Change: Inevitable Surprises*, National Research Council Committee on Abrupt Climate Change, Washington, D.C.: National Academies Press (2002).

The quote from Wallace Broecker is drawn from his article "Thermohaline Circulation, the Achilles' Heel of Our Climate System: Will Man-Made CO_2 Upset the Current Balance?" *Science*, vol. 278 (1997).

The effects of the Little Ice Age in Iceland are described in H. H. Lamb, *Climate, History and the Modern World*, second edition (New York: Routledge, 1995).

The voluminous findings of the Arctic Climate Impact Assessment have been summarized in *Impacts of a Warming Arctic: Arctic Climate Impact Assessment* (Cambridge: Cambridge University Press, 2004).

Chapter 4: The Butterfly and the Toad

The most complete and up-to-date source on the habits and whereabouts of British butterflies is Jim Asher et al., *The*

Millennium Atlas of Butterflies in Britain and Ireland (Oxford: Oxford University Press, 2001).

The Victorians' passion for butterflies is documented in Michael A. Salmon, *The Aurelian Legacy: British Butterflies and Their Collectors* (Berkeley: University of California Press, 2000).

The range changes of European butterflies are described in Camille Parmesan et al., "Poleward Shifts in Geographical Ranges of Butterfly Species Associated with Regional Warming," *Nature*, vol. 399 (1999).

Information on the mating habits of frogs in upstate New York is drawn from J. Gibbs and A. Breisch, "Climate Warming and Calling Phenology of Frogs near Ithaca, New York, 1900–1999," *Conservation Biology*, vol. 15 (2001); on flowering times at the Arnold Arboretum from Daniel Primack et al., "Herbarium Specimens Demonstrate Earlier Flowering Times in Response to Warming in Boston," *American Journal of Botany*, vol. 91 (2004); on Costa Rican birds from J. Alan Pounds et al., "Biological Response to Climate Change on a Tropical Mountain," *Nature*, vol. 398 (1999); on Alpine plants from Georg Grabherr et al., "Climate Effects on Mountain Plants," *Nature*, vol. 368 (1994); and on the Edith's Checkerspot butterfly from Camille Parmesan, "Climate and Species Range," *Nature*, vol. 382 (1996).

A useful book on the biological impacts of warming is Thomas E. Lovejoy and Lee Hannah, editors, *Climate Change and Biodiversity* (New Haven: Yale University Press, 2005).

William Bradshaw published his study of *Wyeomyia smithii* living at different elevations in *Nature*, vol. 262 (1976). The evolutionary effects of climate change are documented in William E. Bradshaw and Christina M. Holzapfel, "Genetic Shift in Photoperiod Response Correlated with Global Warming," *Proceedings of the National Academy of Sciences*, vol. 98 (2001).

Jay Savage's description of discovering the golden toad, as well as a thorough description of the biology of Monteverde, can be found in Nalini M. Nadkarni and Nathaniel T. Wheelwright, editors, *Monteverde: Ecology and Conservation of a Tropical Cloud Forest* (New York: Oxford University Press, 2000). Details of the golden toad's life cycle are drawn from Jay M. Savage, *The Amphibians and Reptiles of Costa Rica: A Herpetofauna Between Two Continents, Between Two Seas* (Chicago: University of Chicago Press, 2002).

The demise of the golden toad is linked to precipitation patterns in J. Alan Pounds et al., "Biological Response to Climate Change on a Tropical Mountain," *Nature*, vol. 398 (1999). Efforts to model the future of the cloud forest are detailed in Christopher J. Still et al., "Simulating the Effects of Climate Change on Tropical Montane Cloud Forests," *Nature*, vol. 398 (1999).

The quotes from G. Russell Coope are drawn from his essay "The Palaeoclimatological Significance of Late Cenozoic Coleoptera: Familiar Species in Very Unfamiliar Circumstances," which appeared in Stephen J. Culver and Peter F. Rawson, editors, *Biotic Response to Global Change: The Last 145 Million Years* (Cambridge: Cambridge University Press, 2000).

The figures on potential extinctions are drawn from C. D. Thomas et al., "Extinction Risk from Climate Change," *Nature*, vol. 427 (2004).

Chapter 5: The Curse of Akkad

An introduction to Akkadian civilization can be found in Marc Van De Mieroop, *A History of the Ancient Near East* (Malden, MA: Blackwell Publishing, 2004).

The verses from *The Curse of Akkad* come from Jerrold S. Cooper, *The Curse of Agade* (Baltimore: The Johns Hopkins University Press, 1983).

A detailed description of Tell Leilan can be found in the chapter on the site by Harvey Weiss in *The Oxford Encyclopedia of Archaeology in the Near East*, vol. 3, Eric M. Meyers, editor (Oxford: Oxford University Press, 1997). Climate change was first proposed as the cause of the abandonment of Tell Leilan in Harvey Weiss et al., "The Genesis and Collapse of Third Millennium North Mesopotamian Civilization," *Science*, vol. 261 (1993).

An overview of the connections between climate change and societal collapse can be found in Peter B. deMenocal, "Cultural Responses to Climate Change During the Late Holocene," *Science*, vol. 292 (2001).

The findings of American scientists in Lake Chichancanab are described in David Hodell et al., "Possible Role of Climate in the Collapse of Classic Maya Civilization," *Nature*, vol. 375 (1995). The findings of researchers off the coast of Venezuela are described in Gerald Haug et al., "Climate and the Collapse of Mayan Civilization," *Science*, vol. 299 (2003). A detailed discussion of drought and Mayan civilization can be found in Richardson B. Gill, *The Great Maya Droughts: Water, Life, and Death* (Albuquerque: University of New Mexico Press, 2001).

James Hansen spoke of being "captivated" by the study of greenhouse warming in a speech titled "Dangerous Anthropogenic Interference: A Discussion of Humanity's Faustian Climate Bargain and the Payments Coming Due," delivered at the University of Iowa, October 26, 2004.

Predictions of warming-induced water shortages in the United States come from David Rind et al., "Potential Evapotranspiration and the Likelihood of Future Drought," *Journal of Geophysical Research*, vol. 95 (1990).

Peter deMenocal wrote about the connections between climate change and human evolution in "African Climate Change and Faunal Evolution During the Pliocene-Pleistocene," *Earth and Planetary Science Letters*, vol. 220 (2004).

Evidence of the drought in Tell Leilan in sediments from the Gulf of Oman is presented in Heidi Cullen et al., "Climate Change and Collapse of the Akkadian Empire: Evidence from the Deep Sea," *Geology*, vol. 28 (2000).

The connection between climate change and the disintegration of Harappan society is discussed in M. Staubwasser et al., "Climate Change at the 4.2 Ka BP Termination of the Indus Valley Civilization and Holocene South Asian Monsoon Variability," *Geophysical Research Letters*, vol. 30, no. 8 (2003).

Chapter 6: Floating Houses

Figures on the Netherlands' water management system come from the Dutch Ministry of Transport, Public Works and Water Management, *Water in the Netherlands: 2004–2005* (The Hague, 2004).

Figures on sea level rise are drawn from the Intergovernmental Panel on Climate Change, *Climate Change 2001: The Scientific Basis*, J. T. Houghton et al., editors (Cambridge: Cambridge University Press, 2001).

The study of flooding commissioned by the British government is cited in David A. King, "Climate Change Science: Adapt, Mitigate, or Ignore?" *Science*, vol. 303 (2004).

A detailed analysis of the Vostok core can be found in Jean Robert Petit et al., "Climate and Atmospheric History of the Past 420,000 Years from the Vostok Ice Core, Antarctica," *Nature*, vol. 399 (1999).

Discussions of the threshold for "dangerous anthropogenic interference" can be found in Brian C. O'Neill and Michael Oppenheimer, "Dangerous Climate Impacts and the Kyoto Protocol, *Science*, vol. 296 (2002); and James Hansen, "A Slippery Slope: How Much Global Warming Constitutes 'Dangerous Anthropogenic Interference'?" *Climatic Change*, vol. 68 (2005).

Chapter 7: Business as Usual

The Environmental Protection Agency's personal emissions calculator can be found at: http://yosemite.epa.gov/oar/globalwarming.nsf/content/ResourceCenterToolsGHGCalculator.html.

Stephen Pacala and Robert Socolow laid out their "wedge" scheme in "Stabilization Wedges: Solving the Climate Problem for the Next 50 Years with Current Technologies," *Science*, vol. 305 (2004).

Information on the fuel efficiency of U.S. autos comes from the report "Light-Duty Automotive Technology and Fuel Economy Trends," Advanced Technology Division, Office of Transportation and Air Quality, U.S. Environmental Protection Agency, July 2005.

The need for new energy technologies to stabilize CO_2 is discussed in Martin Hoffert et al., "Advanced Technology Paths to Global Climate Stability: Energy for a Greenhouse Planet," *Science*, vol. 298 (2002); and also in Hoffert et al., "Energy Implications of Future Stabilization of Atmospheric CO_2 Content," *Nature*, vol. 395 (1998). Martin Hoffert and Seth Potter wrote about space-based solar power in "Beam It Down: How the New Satellites Can Power the World," *Technology Review*, October 1, 1997.

Chapter 8: The Day After Kyoto

Former treasury secretary Paul O'Neill's speculations about Vice President Dick Cheney are recounted in Ron Suskind, *The Price of Loyalty: George W. Bush, the White House, and the Education of Paul O'Neill* (New York: Simon & Schuster, 2004).

The high degree of consensus on global warming is documented by Naomi Oreskes, "The Scientific Consensus on Climate Change," *Science*, vol. 306 (2004).

The Bush administration's editing of climate science reports was disclosed by Andrew C. Revkin, "Bush Aide Edited Climate Reports," *New York Times*, June 8, 2005.

The Bush administration's efforts to water down the proposal for joint action at the 2005 G8 Summit were reported by Juliet Eilperin, "U.S. Pressure Weakens G8 Climate Plan," *Washington Post*, June 17, 2005.

Chapter 10: Man in the Anthropocene

Paul J. Crutzen wrote about the dawn of the Anthropocene and the "luck" that spared the world from catastrophic ozone loss in his essay "Geology of Mankind," *Nature*, vol. 415 (2002).

Sherwood Rowland recounts his reaction to his discovery in Heather Newbold, editor, *Life Stories: World-Renowned Scientists Reflect on Their Lives and the Future of Life on Earth* (Berkeley: University of California Press, 2000).

The discovery of the "ozone hole" is related in Stephen O. Anderson and K. Madhava Sarma, *Protecting the Ozone Layer: The United Nations History* (London/Sterling, VA: Earthscan Publications, 2002).

The amount of warming still required to bring the earth into energy balance is discussed in James Hansen et al., "Earth's Energy Imbalance: Confirmation and Implications," *Science*, vol. 308 (2005).

Afterword

Model predictions of hurricane intensity come from Thomas R. Knutson and Robert E. Tuleya, "Impact of CO_2–Induced Warming on Simulated Hurricane Intensity and Precipitation: Sensitivity to the Choice of Climate Model and Convective Parameterization," *Journal of Climate*, vol. 17, no. 18 (2004).

Kerry Emanuel reported his findings in his article "Increasing Destructiveness of Tropical Cyclones over the Past 30 Years," *Nature*, vol. 436 (2005).

The Georgia Tech group reported its findings in P. J. Webster et al., "Changes in Tropical Cyclone Number, Duration, and Intensity in a Warming Environment," *Science*, vol. 309 (2005).

The effects of rising CO_2 levels on ocean life are discussed in James C. Orr et al., "Anthropogenic Ocean Acidification over the Twenty-first Century and Its Impact on Calcifying Organisms," *Nature*, vol. 437 (2005).

The doubling of Greenland ice loss was reported in Eric Rignot and Pannir Kanagaratnam, "Changes in the Velocity Structure of the Greenland Ice Sheet," *Science*, vol. 311 (2006).

Satellite measurements of Antarctic ice loss come from Isabella Velicogna and John Wahr, "Measurements of Time-Variable

Gravity Show Mass Loss in Antarctica," *Science Express*, March 2, 2006.

The Bush administration's efforts to censor James Hansen were first reported by Andrew C. Revkin, "Climate Expert Says NASA Tried to Silence Him," *New York Times*, January 29, 2006.

Bush's meeting with Michael Crichton is recounted by Fred Barnes, *Rebel-in-Chief: Inside the Bold and Controversial Presidency of George W. Bush* (New York: Crown Forum, 2006).

Al Gore described himself as "optimistic" about the U.S. response to climate change on *Fresh Air*, National Public Radio, May 30, 2006.

RESOURCES

Hundreds of groups in the U.S. are working to curb greenhouse gas emissions on the local, state, and national levels. Many provide information on their Web sites about how individuals can reduce their own "carbon footprints." Others offer updates on scientific and political news.

The Cities for Climate Protection Campaign offers assistance to local governments looking to cut emissions. Information on the campaign can be found at www.iclei.org.

Religious groups working to curb global warming include:
The National Religious Partnership for the Environment:
www.npre.org
The Coalition on the Environment and Jewish Life: www.coejl.org
Evangelical Environmental Network: www.creationcare.org.
The Regeneration Project: www.theregenerationproject.org

The Pew Center on Climate Change works with business and policy leaders: www.pewclimate.org.

A number of regional groups are working to cut emissions. These include:
Climate Solutions: www.climatesolutions.org
The Southern Alliance for Clean Energy: www.cleanenergy.org
The New England Climate Coalition: www.newenglandclimate.org

The Rocky Mountain Climate Organization:
www.rockymountainclimate.org

News, energy-saving advice, and a "carbon calculator" can be found at www.stopglobalwarming.org.

RealClimate provides explanations of and commentary on the latest climate studies by leading scientists: www.realclimate.org. More climate science news can be found at www.giss.nasa.gov.

The National Environmental Trust follows the latest legislative developments on global warming: www.net.org.

Many environmental groups offer information on global warming on their Web sites. These include:
Natural Resources Defense Council: www.nrdc.org
Environmental Defense: www.environmentaldefense.org
Greenpeace: www.greenpeace.org
The Sierra Club: www.sierraclub.org
Friends of the Earth: www.foe.org
The Union of Concerned Scientists: www.ucsusa.org

The American Council for an Energy-Efficient Economy offers detailed information on energy-efficient appliances, including dishwashers, refrigerators, and air-conditioning systems: www.aceee.org. The group also publishes an automotive guide at www.greenercars.com.

If you are interested in installing solar panels on your home or business, www.findsolar.com can help you to locate a licensed installer in your area.

Many utilities offer customers energy audits of their homes. Some also provide rebates for the purchase of energy-efficient appliances and offer consumers the option of purchasing "green" power (usually at a premium). Check with your local utility.

INDEX

Note: Information presented in figures is denoted by *f*.

A NOTE ON THE TYPE

The text of this book is set in Bembo. This type was first used in 1495 by the Venetian printer Aldus Manutius for Cardinal Bembo's *De Aetna*, and was cut for Manutius by Francesco Griffo. It was one of the types used by Claude Garamond (1480–1561) as a model for his Romain de L'Université, and so it was the forerunner of what became standard European type for the following two centuries. Its modern form follows the original types and was designed for Monotype in 1929.